Heart of Carbon

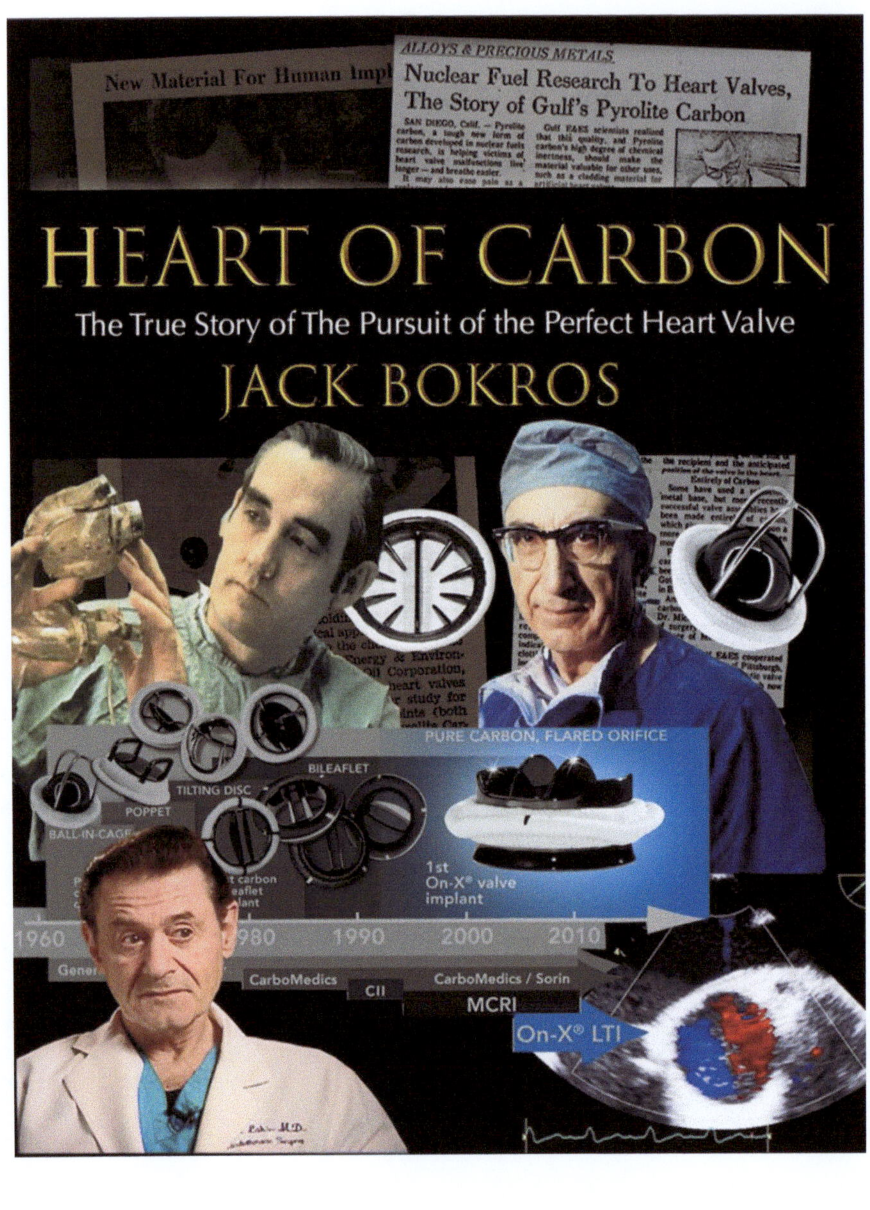
ALLOYS & PRECIOUS METALS
New Material For Human Implant
Nuclear Fuel Research To Heart Valves,
The Story of Gulf's Pyrolite Carbon

HEART OF CARBON
The True Story of The Pursuit of the Perfect Heart Valve
JACK BOKROS

PURE CARBON, FLARED ORIFICE
BILEAFLET
TILTING DISC
POPPET
BALL-IN-CAGE
1st On-X® valve implant
1960
1980
1990
2000
2010
CarboMedics
CII
CarboMedics / Sorin
MCRI
On-X® LTI

Jack Bokros

Heart of Carbon

The Story Behind the Pursuit of the Perfect Mechanical Heart Valve

 Springer

Jack Bokros
Georgetown, TX, USA

This work contains media enhancements, which are displayed with a "play" icon. Material in the print book can be viewed on a mobile device by downloading the Springer Nature "More Media" app available in the major app stores. The media enhancements in the online version of the work can be accessed directly by authorized users.

ISBN 978-3-031-17935-8   ISBN 978-3-031-17933-4   (eBook)
https://doi.org/10.1007/978-3-031-17933-4

Cover Illustration: Michael Markl, PhD, Northwestern University Feinberg School of Medicine

This Springer imprint is published by the registered company Springer Nature Switzerland AG
The registered company address is: Gewerbestrasse 11, 6330 Cham, Switzerland

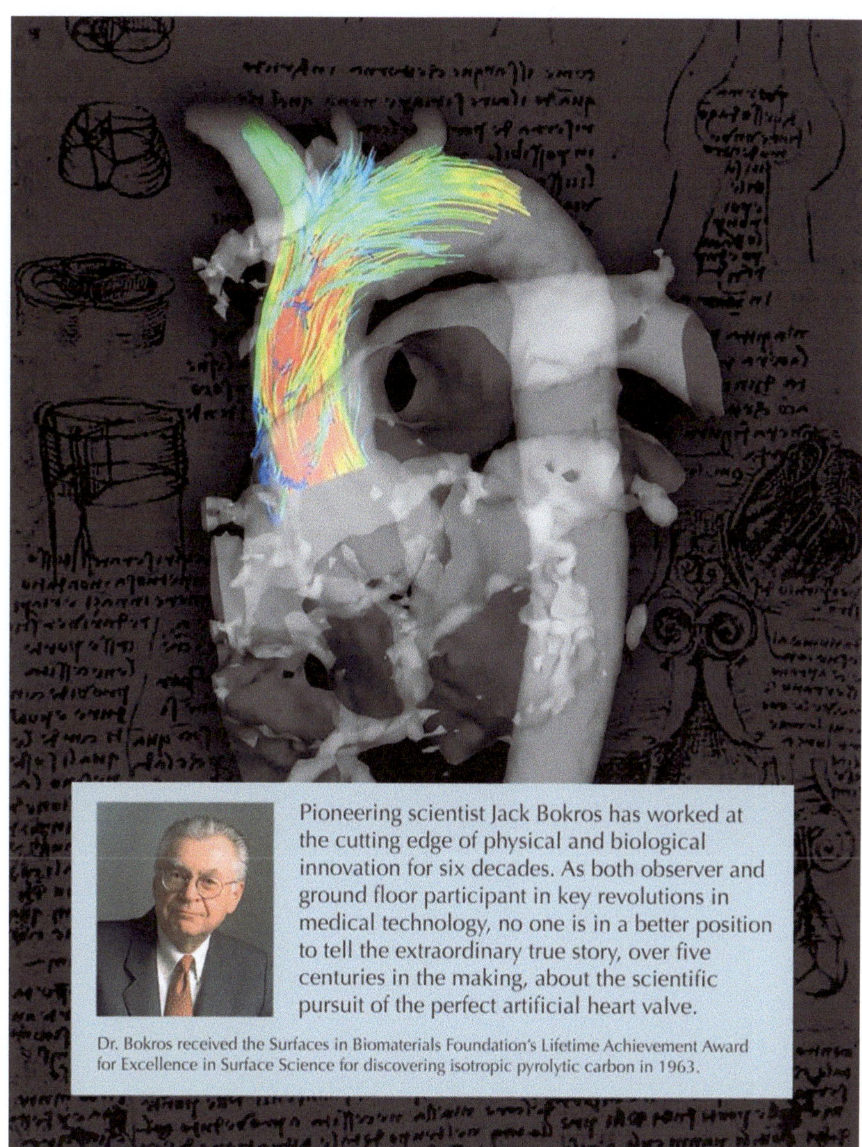

Pioneering scientist Jack Bokros has worked at the cutting edge of physical and biological innovation for six decades. As both observer and ground floor participant in key revolutions in medical technology, no one is in a better position to tell the extraordinary true story, over five centuries in the making, about the scientific pursuit of the perfect artificial heart valve.

Dr. Bokros received the Surfaces in Biomaterials Foundation's Lifetime Achievement Award for Excellence in Surface Science for discovering isotropic pyrolytic carbon in 1963.

Disclaimer

Use of copyrighted content herein owned by Artivion, Inc. or its affiliates does not constitute or imply agreement with or endorsement of the views expressed herein, which are the author's own.

Foreword by John D. Puskas, MD

I was first introduced to the On-X mechanical valve by my colleague, Dr. Omar Lattouf, shortly after it was approved for use in the aortic position by the FDA in 2001. Omar had previously used it while operating in his native Jordan and explained to me its obvious design advantages. I immediately tried it, was impressed by the very low gradients obtained, and adopted it as my routine choice for mechanical aortic valve replacement. Since FDA approval of the On-X mitral valve in 2002, I have never implanted a mechanical valve that was not an On-X valve.

John Ely, VP for Regulatory Affairs at Medical Carbon Research Institute (manufacturers of the On-X valve), invited me to have breakfast with him during a national cardiothoracic surgical meeting in 2005. At that breakfast, he explained the data from Europe and South Africa that made him believe that the On-X valve might be safely used with lower levels of warfarin anticoagulation than were mandated by contemporary guidelines. When asked whether I would consider leading such a trial, I agreed. At that breakfast, literally on a napkin, the design of the PROACT trial was first drafted. I asked that two important features be included in the trial design: first, that all patients should receive standard warfarin anticoagulation plus low-dose aspirin for three postoperative months to allow endothelialization of the sewing cuff prior to randomization; and second, that we include a group of low-risk AVR patients who would then be randomized to receive dual antiplatelet therapy alone, without warfarin. As is well known, the PROACT trial resulted in FDA approval of an Indication For Use (IFU) for a lower INR target range (INR 1.5–2.0 plus low-dose aspirin) than the traditional INR target of 2.0–3.0 plus low-dose aspirin for patients implanted with an On-X AVR, after 3 postoperative months of traditional anticoagulation. While the dual antiplatelet therapy arm failed to demonstrate non-inferiority, results of reduced INR in patients with the On-X mechanical aortic valve showed a striking reduction in bleeding events without an increase in thromboembolic events and led to the first and only FDA IFU approval for lower INR for a mechanical heart valve prosthesis. Importantly, at the time of this writing, analysis of the results of lower INR (INR 2.0–2.5) for patients with an On-X valve in the *mitral* position is underway and preliminary results seem favorable for an upcoming FDA submission.

The historic ruling by FDA to allow a lower INR with (only) the On-X mechanical heart valve is an affirmation of the natural laminar blood flow through the On-X valve due to its flared orifice design as well as the thromboresistant polished surfaces of its pure carbon. There are other design features that bear mention, namely the supra-annular sewing cuff, the cylindrical housing in which the valve leaflets close with remarkably small regurgitant volumes and low leaflet impact velocities, and the actuated pivots with large purge volumes that eliminate blood stasis, reducing the overall thrombogenicity of the valve prosthesis. All of these design features are the work of Jack Bokros, PhD, a remarkable materials scientist and engineer who also played a key role in the design of the St. Jude mechanical heart valve prosthesis decades earlier.

In this book, Dr. Bokros provides a personal account of his own pioneering work as his career evolved over four decades from engineering nuclear reactors and carbon materials in the aftermath of the Manhattan Project to his invention of the On-X mechanical heart valve. This saga included multiple twists and turns worthy of a television drama series, replete with corporate espionage, international intrigue during the Cold War, wrongdoing and litigation, groundbreaking scientific achievement, and relentless determination to finally create a pure carbon heart valve through which blood flow is near-normal. This chronicle reminds one of the statement by Thomas Edison that genius is 1% inspiration and 99% perspiration. It is also another example in which progress in one field (engineered carbon for nuclear reactors) led to a breakthrough in an unrelated field (surgical treatment of valvular heart disease). Taken together, these dual lessons should teach aspiring cardiovascular innovators to leverage advances in allied fields and to be prepared for a long, arduous pathway to success.

Several of the early giants of cardiac surgery served important supporting roles for the inventor-protagonist. But such a story could only be told by the architect of the adventure himself. In this very personal book, Dr. Bokros demonstrates that he is not just a pioneering scientist but also a captivating storyteller.

John D. Puskas
National Principal Investigator
Prospective Randomized On-X Anticoagulation Clinical Trial (PROACT)
New York City, NY, USA

International Principal Investigator
PROACT Trial
New York City, NY, USA

Chairman, Department of Cardiovascular Surgery, Mount Sinai Morningside
Mount Sinai West and Mount Sinai Downtown Hospitals
New York City, NY, USA

Professor of Cardiovascular Surgery
Icahn School of Medicine at Mount Sinai
New York City, NY, USA

Foreword by Mervyn Williams

Jack Bokros is a legendary figure in the development of prosthetic heart valves. When Jack asked me to write a preface for his book, I felt honored. But then I became hesitant because much of what I would have to say would be about myself, being one of the early users and protagonists of the On-X valve. On reflection, I decided that documenting my experience with valve replacement was relevant since, like Jack's experience, it begins with the industrially inspired design of the caged ball valves and ends with the elegance of the On-X valve. Jack and I have made similar journeys during the development of mechanical heart valves, albeit on different roads—mine in the operating theater, his in the laboratory.

In 1966, I commenced my training as a cardiothoracic surgeon. At that time, heart valve replacement was at the beginning of what was to turn out to be a very long learning curve. The first successful mitral valve replacement had been carried out by Nina Braunwald, and the first aortic valve replacement by Dwight Harken in 1960. Remarkably, Harken's patient was still alive 20 years later. Harken set out the four important requirements for a prosthetic heart valve: *easy* to implant, *resistant* to thrombotic events, *hemodynamics* approaching those of the native valve, and *durability*. These aims are still relevant, and the history of valve replacement surgery is essentially the pursuit of these targets.

The first valve that I implanted was a mitral valve in a 16 year old boy with rheumatic valve disease. The valve implanted was a Starr Edwards caged ball device, the only valve commercially available at the time. A month after his discharge, he was readmitted with a massive cerebral embolus and died. Whether or not he took the prescribed anticoagulants is not known. For a surgeon at the beginning of his career, this was a shattering but not unexpected event. Most patients needing valve surgery in South Africa come from the historically disadvantaged segment of the population. Poorly educated and living in crowded poverty-stricken communities, the management of anticoagulation is difficult and often impossible. Unsurprisingly, thrombosis, thromboembolism, and bleeding are not uncommon after valve surgery.

When I established the cardiac unit in Port Elizabeth, the valves we used initially were the St. Jude and the Hall-Kaster (later the Medtronic Hall valve). The St. Jude valve was developed and manufactured by Jack's company. It was to become the most successful heart valve in history with 2 million implants by 2011. A full account of the development of the St. Jude valve will be found in Jack's book. The disc used in the Medtronic Hall valve was also manufactured by Jack, as were the Pyrolite Carbon components of most valves available at the time.

Thrombotic obstruction was encountered in both of the valves we used, but the mechanism was different. In the St. Jude valve, thrombosis was in the hinge mechanism. In the Medtronic Hall valve, thrombosis was related to pannus ingrowth. Overall, thrombus was lower in the Medtronic Hall valve, and it became our valve of choice. Our experience was mirrored by Manuel Antunes in Johannesburg, his patients coming from a population similar to ours. He published a series of 2000 patients before moving to Coimbra, Portugal, where he continued to use the Medtronic Hall valve. He was devastated when Medtronic stopped making the valve and bought all the remaining stock for future use!

By the early 1990s, we had accumulated a collection of explanted valves obstructed by pannus, and examining them I wondered whether pannus could be prevented from obstructing the valve disc by placing the disc in a tubular housing. I approached Medtronic, and they arranged for animal experiments at the NIH to explore this concept. The implanted valves with tubular housing showed that pannus, while it still occurred, did not encroach on the orifice. I asked them to manufacture a valve with a tubular rather than the existing washer-like housing. They replied that they were primarily a pacemaker company and were not prepared to embark on a major redesign of the valve. I created a few slides illustrating my concept and presented them whenever I lectured on valve surgery.

Then, in 1999, paging through a medical journal, I came across a full page advertisement for the On-X valve. I noted that the valve leaflets were housed in a cylinder. I was elated—at last my valve had arrived! I contacted the South African distributor who informed me that the valve was made by MCRI, a small company founded by Jack Bokros, and that Jack would be in London at a congress in a few weeks' time. I had planned to attend the same congress and arranged to meet him there. The distributor told me that only one valve, an aortic valve, had been implanted in South Africa. The surgeon was Chris Hammond. I knew Chris well and contacted him. He told me it was a "lekker" valve. Lekker is an Afrikaans word roughly translated as desirable. He also said that a scalloped sewing ring would make implantation easier. Many years later, On-X added a scalloped sewing to their options. Sometimes it takes a long time for a good idea to be recognized!

Although Jack was part of the history of heart valve development, and I was well aware of the massive impact his introduction of Pyrolite Carbon had made in the manufacture of heart valves, I had never met him. We met in the foyer of his hotel in London. He was accompanied by John Ely. I told them that the cylindrical housing echoed a concept that I had promoted whenever I had the opportunity, and that I would be interested in implanting his valve and comparing the results with data we had collected over the last 15 years using other valves. They proceeded to point out the other features that gave On-X advantages over competing valves. By eliminating silicon from the manufacturing process, the resulting pyrolytic carbon, which they called On-X carbon, was smoother and much more versatile, allowing innovative design features, notably design of the hinge mechanism. Other advantages are detailed in his book. One feature that bothered me was that the leaflets were parallel when fully open. Medtronic had produced a valve called the Parallel Valve. The valve was withdrawn. I was reassured that the hinge mechanism of the On-X was completely different and that thrombosis had not been reported. Also, the valve was to be described as opening to 90°.

We implanted the first valve in 1999, and to date, more than 4000 valves have been implanted in South Africa, mostly in noncompliant patients. We published early and midterm results. Our 15-year results are to be found in this book. Valve thrombosis has not been eliminated but is uncommon. Pannus has not been encountered in our patients, nor has it been reported elsewhere. There is no doubt that this is due to the tubular configuration of the housing.

Since our first meeting in London, I have seen a lot of Jack Bokros, and we have become good friends. If I had to choose one word to describe Jack, I would choose enthusiasm. At the Asian meeting a few years ago, his enthusiasm was concentrated on the Italian polymath Leonardo Da Vinci. Da Vinci's drawings of flow across the aortic valve are strikingly similar to the flow across the On-X valve. His enthusiasm for Da Vinci has prompted a half hour video, the leading actors being the On-X valve and the Italian Leonardo. Over the years, I have been entertained by Jack's reminiscences of his experiences at the coalface of valve development (this may be a pun) and reading his book has given me the pleasure of remembering his stories. The On-X valve is currently the best valve available but is not perfect. Although Jack recently turned 90, I have no doubt that he is still thinking up further improvements!

Mervyn Williams
Chief Surgeon and Head of Department
of Cardiothoracic Surgery
Provincial Hospital,
Port Elizabeth, South Africa

Acknowledgments

As the *Heart of Carbon* book extends over six decades, persons making extraordinary contributions are numerous and extend over a number of phases.

In the beginning (1953), *Walter P. Wallace*, at the Hanford Manhattan site, charged me with a problem that was solved during my 3-month intern period. Ultimately, that effort led to a staff member position at General Atomic. In 1962, *David Ragone* provided seed money to explore other uses for isotropic pyrolytic carbon and found it to be suitable for the construction of heart valve replacements—a pivotal game changer. *Robert Akins, Axel Haubold, John Sommerfeld*, and *Michael Emken*, the team that responded to *Harry Cromie's* request for carbon valves for his surgeons, that is, *Vincent Gott, Michael DeBakey,* and *Arthur Beall,* all became pioneering contributors.

In the mid-1980s, *Rolland Siegel* joined CarboMedics to spearhead worldwide marketing of the CarboMedics valve and later, in the Medical Carbon Research Institute's On-X valve marketing efforts, became a crucial and enduring contributor.

At Intermedics and CarboMedics, *GR Chambers* and his son *Russell Chambers*, with *Lou Marinos* and *Jerry Klawitter,* perpetuated the carbon story. Their efforts are much appreciated. In the realm of developing carbon heart valve replacements, *Robert (Bob) Kaster* and *Dr. "Walt" Lillehei* were major contributors that led to the development of the St. Jude valve by *Manny Villafana*. Perhaps most important were *John Ely's* contributions, first to Hemex and the Medical Carbon Research Institute, and then on to the On-X valve development.

In the mid-1980s, *George Sines*, a renowned expert in fracture mechanics at UCLA, chose *Ling Ma* to study and quantify the fundamental mechanical properties of pyrolytic carbon and to produce a thesis. The summary of Ma's findings describes basic aspects of the science and technology of carbon and are reproduced in the appendix.

In 1985, Professor *Ned HC Hwang*, an expert in fluid mechanics, joined CarboMedics to develop a valve for China. He, along with *Luo Zheng Xian*, a surgeon and colleague of Hwang's and head of the Guangdong Hospital, *Zhang Ho*, and *Cai Yongzhi* (who implanted the first homemade ball valve in China) are all major contributors. In the same timeframe, *Rolland Siegel* chose Professor *J Y Neveux* to implant the first CarboMedics valve in France, and then he and Neveux traveled to China in support of marketing efforts there.

In 1989, *Robert Akins* retired and appointed *Jonathan Stupka* to replace him, both major contributors. In 1992, working on the Medtronic Project, *David Wilde* and *Jim Accuntius* engineered a paradigm shift in carbon processing that constituted a giant step toward developing an ideal valve replacement.

When Medical Carbon Research Institute was formed in 1994, Russell Chambers suggested two surgeons—Professor *Ernst Wolner* and *Francis Fontan*—as consultants. Both were icons in medicine and were instrumental in making the On-X valve a reality.

In 1996, Professor *Axel Laczkowits* performed the first On-X valve implant, launching the On-X phase, a major contribution that led up to *Mervyn Williams* in 1999 initiating his 15-year clinical trial in South Africa. *Sidney Levitsky* volunteered to chair the Safety Monitoring Committee for the Trial, an important contribution.

In 2004, *Hillel Laks* was the first to confirm the On-X aortic valve's ability to preserve the normal native blood flow—a major revelation.

In the same year, with a broad background and highly recommended, *Sherry ShawSmith* agreed to join Medical Carbon Research Institute and almost immediately (she had already checked us out) declared that the "carbon story" should be told. She enlisted *Vincent Gott* to promote publishing a book and, with great energy, took the lead in marketing and sales, with the mantra "How's the book going?" continually ringing in my ear: truly inspirational. In the last two decades, she was a prime mover and much appreciated. Thank you, Sherry.

John Puskas became the Principal Investigator of the PROACT in 2005 continuing to FDA pre-market approval for use of a low level intensity of anticoagulation in 2015, an outstanding accomplishment.

Beginning in the 1980s, *Kathleen Selbrede* made important contributions to marketing as a graphic artist and later with her husband, *Martin Selbrede*, took on the production of a video. The Selbredes, working with *Gerald Buckberg, Philip Kilner,* and *Michael Markl*, produced the Leonardo da Vinci video, a remarkable and much appreciated contribution. Kathleen's creation of animated images of blood flow through the On-X valve derived from a 500 year old manuscript of Leonardo da Vinci is spectacular.

Gerald Buckberg and *Endre Bodnar*, both icons in medical science, were appreciated mentors.

Contents

Abbreviations

3M	Minnesota Mining and Manufacturing Company
4D-MRI	Four-Dimensional Magnetic Resonance Imaging
AHP	Artificial Heart Program
ARR	Aortic Root Replacement
ASAP	As Soon As Possible
ATS	Advancing the Standard Medical
AVR	Aortic Valve Replacement
BSCC	Björk Shiley Convexo-Concave
CEO	Chief Executive Officer
CFO	Chief Financial Officer
CMI	CarboMedics, Inc.
COO	Chief Operating Officer
CV	Cardiovascular
CVS	Cardiovascular Surgery
DMed	Doctor of Medicine
DNA	Deoxyribonucleic Acid
EACTS	European Association for Cardio-Thoracic Surgery
EKG	Electrocardiogram
FDA	Food and Drug Administration
GBH	Graphite Benzalkonium Chloride-Heparin
HTGR	High Temperature Gas-Cooled Reactor
HVMB	Helical Ventricular Myocardial Band
IDE	Investigational Device Exemption
INR	International Normalized Ratio
LDH	Lactate Dehydrogenase
LLC	Limited Liability Company
LTI	Isotropic carbon deposited at low temperature near 1300° C
MCRI	Medical Carbon Research Institute, LLC
NIH	National Institutes of Health
OEM	Original Equipment Manufacturer
On-XLTI	On-X Life Technologies, Inc.

PMA	Pre-Market Approval
PRC	People's Republic of China
PROACT	Prospective Randomized On-X Anticoagulation Clinical Trial
rAVR	Robotic Aortic Valve Replacement
R&D	Research and Development
RBC	Red Blood Cells
RBV	Regional Backflow Velocity
SEM	Scanning Electron Microscopy
SiC	Silicon Carbide
SVD	Structural Valve Deterioration
TAVI	Transcatheter Aortic Valve Implantation
TRIGA	Training, Research, Isotopes, General Atomics
UCLA	University of California at Los Angeles
USA	United States of America
US-FDA	United States Food and Drug Administration
VC-CEO	Venture Capitalist-Chief Executive Officer
VHE	Vorticity Helicity Eccentricity
VP	Vice President
WSS	Wall Shear Stresses

List of Figures

List of Tables

Chapter 1
Overview

As our timeline (Fig. 1.1) shows, the carbon story began in 1953. I was a student intern at the Manhattan Project's Hanford site, where plutonium was being produced for bombs. My boss, PhD metallurgist Walter Wallace, charged me with solving a difficult problem that had plagued the project since the inception of plutonium production efforts. Three months later, I happened upon the solution. Dr. Wallace was elated.

I later graduated from the University of Wisconsin and focused on my career objective: the generation of electricity from nuclear power. Dr. Wallace, who'd remembered my earlier contribution to the Manhattan Project, called me in 1958. He had been selected by General Atomic to support its effort to develop a high-temperature gas-cooled reactor (the HTGR). While the massive containment vessel of the reactor was made of steel, its core depended on graphite and carbon-coated uranium-thorium carbide particles.

Dr. Wallace asked me to join him at General Atomic as a staff member. His recommendation on my behalf must have carried considerable weight. The majority of staff members were scientists from the Manhattan Project or professors in their specific fields and traveled first class.

My superior was Professor David Ragone, on leave from the University of Michigan. He balked when he was told that I did not have a PhD, asserting that all staff members needed their "union card" (a doctoral degree).

Despite new personal obligations arising in 1959 (my marriage to Roberta and the birth of our first daughter, Kathleen), I enrolled at the University of California at Berkeley in the fall of 1960. After 2 years of intensive study, my thesis was accepted in 1962. I was awarded the PhD in Materials Science in 1963.

I called Ragone. Upon hearing my voice, he said, "Where have you been?" After a chat, he came out and said, "I need you to come back and sort out the science of soot."

I knew, of course, that soot was carbonaceous smoke but was puzzled by "the science of soot."

J. Bokros, *Heart of Carbon*, https://doi.org/10.1007/978-3-031-17933-4_1

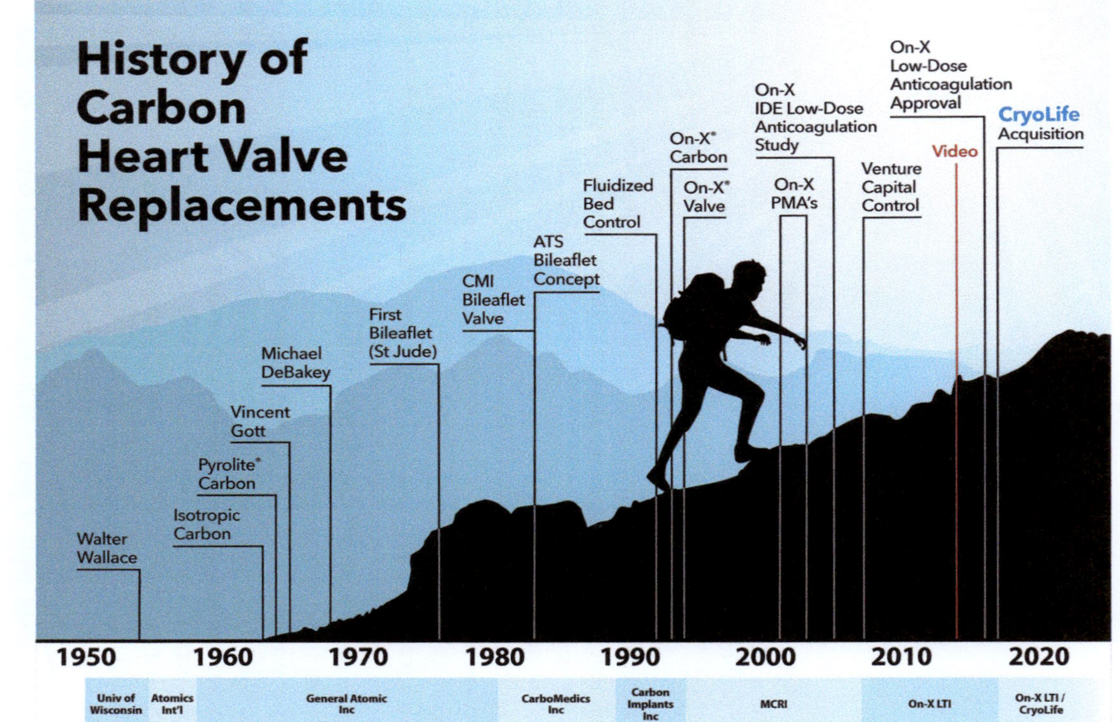

Fig. 1.1 Timeline of events: Discovery of isotropic pyrolytic carbon, DeBakey's first clinical use, first all-carbon bileaflet valve, achievement of bed control, On-X carbon and On-X valve, On-X valve approval by FDA, approval of On-X valve Investigational Device Exemption for low-dose anticoagulation study, venture capital control, On-X valve low-dose anticoagulation FDA approval, and CryoLife acquisition. The sequential entities involved: General Atomic, Inc. (now General Atomics); CarboMedics, Inc.; Carbon Implants, Inc.; Medical Carbon Research Institute, LLC; On-X Life Technologies, Inc.; and CryoLife, Inc. (now Artivion, Inc.). (Reprinted with the permission of Kathleen Selbrede)

A special form of soot had been discovered around 1963: it was isotropic pyrolytic carbon, strong and impermeable. As a coating for nuclear fuel particles, it could contain the fission fragments released by the splitting of uranium atoms, preventing their release into the reactor system.

In the same year isotropic carbon was discovered, I had a chance meeting with Vincent Gott, MD, a cardiac surgeon who had developed a carbon-coated heart valve replacement. He tested our material and discovered that blood did not identify isotropic carbon as a foreign body. Gott introduced us to the National Institutes of Health (NIH) Artificial Heart Program, recommending that the NIH work with General Atomic to help in their artificial heart program.

Frederick de Hoffmann, President of General Atomic, became aware of my unauthorized efforts with Gott which were a diversion from my primary job to develop nuclear fuel. Far from being annoyed, Dr. de Hoffmann was intrigued by this venture into the medical field. He told Ragone to provide support of the medical activity through his $50,000 discretionary fund.

My medical project grew into a self-supporting division of General Atomic, called the Medical Products Division. As the timeline indicates, it flourished from the mid-1960s until the late 1970s.

As an original equipment manufacturer (OEM) supplier of carbon components for use in heart valve replacement, the Medical Products Division became a sole source of Pyrolite® Carbon components to the mechanical heart valve replacement market.

The field was protected by our suite of patents on both materials and production processes. It is an enviable position for a company to be in, but sole sources raise suspicions of anticompetitive conduct. Sensitive to that issue, the oil companies, investors in General Atomic, directed that Medical Products Division avoid antitrust violation by not imposing burdensome pricing or manipulating supply.

The invention of the heart/lung machine by John Heysham Gibbon drove the rapid proliferation of entrepreneurial companies entering the industry. Some promoted concepts bearing the name of notable cardiac surgeons. All sought fame and fortune.

But from the outset, mechanical valve replacement designs simply did not mimic healthy native valves. They did not preserve the natural normal flow of blood within the heart. They served as primitive check valves that opened by systolic pressure and closed at the end of systole. Consequently, the state of the art was quite archaic, controlling the upstream flow but not preserving normal natural streamlined flow. In fact, the importance of normal flow was not appreciated. The elucidation of normal blood flow would have to wait until 2015, when four-dimensional magnetic resonance imaging (4D MRI) finally enabled researchers to visualize and quantitatively characterize the phenomenon.

In 1979, the Medical Products Division of General Atomic was sold to Intermedics, Inc. The transition raised eyebrows with Medical Products Division's customers. While Intermedics was not currently competitive, it posed a potential threat to Medical Products Division's customers whose only source for carbon heart valve components had been swallowed up by a company that *could* compete against them in the future.

The sale was consummated with Intermedics and structured to assure all concerned that the company had no intention of competing with Medical Products Division's customer base. The new entity was called CarboMedics, Inc., and, to insulate proprietary information of its customers from the parent Intermedics, it was set apart geographically. While both were in Texas, CarboMedics, Inc. was located centrally in Austin, while Intermedics was headquartered on the Gulf Coast in Freeport.

The first valve designs used "poppets" (balls or disc-shaped occluders) that moved up- and downstream to selectively restrict flow to side channels. The second series valves were designed to tilt a disc to reduce the opposition to forward flow, but it was still remote from normal natural flow.

The use of two tilting leaflets represented a significant advance in design. Some valves had two leaflets that were centrally located and swung away from central flow. Notable among these were the St. Jude Medical valve and valves designed for the People's Republic of China (PRC). They were the PRC-I and PRC-II by CarboMedics, Inc., as well as a bileaflet valve from an Italian biomedical company. Sales of the latter were originally restricted by CarboMedics, Inc. patents. Other bileaflets (such as designs emanating from Mitral Medical Inc.) used bileaflets that opened naturally away from the forward central flow. The latter never made it to the market due to lack of funding.

Except for the Italian biomedical company's valve, all of the bileaflet valves mentioned above used Pyrolite Carbon components. Those that depended on CarboMedics, Inc. for their components were anxious concerning supply and sought licenses for the sale of General Atomic's proprietary carbons. Those anxieties were determinative in the case of the St. Jude Medical litigation. That litigation triggered a significant change in the CarboMedics business plan. CarboMedics elected to enter the valve market independently via the People's Republic of China with its own in-house designs.

With one exception, all of the mechanical valves up until 1992 were fabricated using components from CarboMedics. In contrast, the Italian biomedical company used a carbon that mimicked the crystalline structure of Pyrolite Carbon. Their valves were restricted to sales outside of General Atomic's proprietary carbons bought by Intermedics and sold through its subsidiary, CarboMedics, Inc.

In 1988, Intermedics was sold to a European company, which paid $800 million for Intermedics, Inc. with about $400 million for the CarboMedics, Inc. group and the balance for the Intermedics pacemaker business. The sale included the CarboMedics group (comprised of CarboMedics, Inc. proper and two entities founded by CarboMedics, Inc.: an orthopedic company and a dental company). At the time, the blanket of General Atomic patents protecting Pyrolite Carbon technology was expiring.

The CarboMedics group management team of five rejected an offer to go with the European company. No longer tethered to the suite of Pyrolite Carbon patents, they sought out other opportunities. Sadly, the plans to set up a Chinese joint venture to buy valve kits that included unfinished Pyrolite Carbon components along with other non-carbon materials did not materialize. The PRC-II design was named the CarboMedics valve and was to be available in China and other countries, mostly in the Far East, but not in the United States or Europe. After the St. Jude Medical litigation, it became available worldwide as the CarboMedics valve.

Before the European deal was consummated, G. Russell Chambers took me aside and told me that he and Wayne Rollins (both major shareholders in Intermedics) would like to provide five million dollars "for whatever venture you do—more could be forthcoming as needed."

Shortly thereafter, the president of Medtronic, Inc.'s Heart Valve Division approached me in San Diego, inviting me to lunch. He wanted a carbon facility and a valve. After going through the European sale disruption, I was in no mood to consider another big company and informed the president that we already had an investor. I had zero interest in a deal with Medtronic, but the president asked for a 90-day hiatus so we could listen to their proposal.

I traveled to Minneapolis without an attorney. An Executive Vice President offered $200,000 for us to hold off for 90 days. I shrugged off the offer. The Executive Vice President read my animated response accurately and answered, "Okay, six-hundred-thousand dollars for the ninety days, but if a deal gets structured, the six-hundred-thousand-dollars would be invested in the venture." I still was not interested and pushed back from the table. Then he said, "Medtronic will fund everything: a carbon facility with a 40,000-valve capacity and a valve they can sell."

And so, the deal was struck. Medtronic would have an option to buy Carbon Implants, a newly formed company, for a substantial sum, and the management team would be free to depart and use the carbon technology for any venture. They called it a Bokros Entity. There were other restrictions imposed on future sales of any Bokros Entity.

The carbon facility was provided along with a bileaflet valve.

The Medtronic agreement demanded that the valve be thoroughly tested by prominent professional experts in their disciplines and ready for clinical trials. The climax for us was a presentation by experts in the field that ranked the valve on a scale of 1–10. The experts ranked our valve at 11.

Confidently, Medtronic announced the design was to be frozen calling it "Design Freeze." Exciting indeed! The valve was labeled "Parallel" because the two leaflets were free to follow the flow and thus be parallel to it without fighting the flow: a unique development in valve design. We were paid in cash with a small holdback for 2 years.

As chance would have it, we made another important improvement in the carbon process during the Carbon Implants project.

From its inception in the 1960s, the carbon process was limited because of a deficiency: the inability to sense one of the parameters (the bed size) during a processing run. In 1992, this deficiency was corrected, liberating design options. The improvement was patented (labeled *Fluidized Bed Control* on the timeline).

Medtronic owned the patent. However, by agreement, we had the right to use the technology without limitation, i.e., as a Bokros Entity.

With cash in hand, we formed the Medical Carbon Research Institute, LLC to use the improved new technology to design a valve (the *On-X® carbon* and *On-X® valve* on the timeline).

Many of the features of the Medtronic, Inc. Parallel Valve were incorporated into the On-X valve. Its length, like that of the Parallel, was near the length of natural valves. The On-X valve leaflet was free to follow the flow to full open. On-X carbon was free of silicon carbide, and the inlet orifice was flared to prevent flow separation. All of its features were fashioned to preserve normal flow, features hitherto unattainable using earlier Pyrolite Carbon.

Shortly after the Parallel Valve was implanted, there were clotting problems that persisted until Medtronic, Inc. abandoned the valve. Even though Medtronic, Inc. owned the patents on the process used to produce On-X carbon, they instead chose Pyrolite Carbon (in their words, a *proven material*) to design the Parallel Valve.

Medical Carbon Research Institute's On-X valve was introduced clinically in 1996 and approved in Europe, earning the CE mark that meant it was in conformity with health, safety, and environmental protection standards within the European Economic Area. This was in 1998. It was later fully approved by the US FDA (aortic in 2001 and mitral in 2002 as labeled on Fig. 1.1).

The FDA trial was supplemented with other studies that enabled the granting of an Investigational Device Exemption (IDE) from FDA approval in 2006. The IDE was to test the On-X valve at reduced levels of anticoagulation.

Our funding from the sale of Carbon Implants to Medtronic, Inc. was running low. An expected stream of royalties from the sales of the Parallel Valve died along with the discontinued valve. Alongside this impediment, our only option was venture capitalists. The venture capitalist appointed a CEO, hereafter the VC-CEO, one inexperienced in valve technology and carbon processing.

After receipt of the IDE in 2006, progress toward approval crawled along at a snail's pace. I was frustrated by the VC-CEO's marketing efforts. Accordingly, late in the approval process, we made an effort (independent of the venture capitalist) to make an educational video (*Video* on the timeline).

When the video was nearly finished in 2015, the VC-CEO advised us that CryoLife, Inc. was interested in On-X Life Technologies (On-XLTI was the entity that formed in 2007 converting Medical Carbon Research Institute from a limited liability company to a C-corporation).

I was relieved to finally have the venture capitalists out of my hair. Moreover, I had known Steve Anderson, the founder of CryoLife, since about 1980. This would be a good fit. The sale to CryoLife was finalized in December of 2015.

I can only speculate as to what our fate would have been had I accepted G. Russell Chamber's proposition in 1989 rather than joining with Medtronic, Inc. While Medtronic, Inc. was a disappointment, the ensuing struggles enabled a world-class heart valve to actually come to market despite corporate setbacks and management myopia.

Per ardua ad astra: through adversity to the stars. This book documents that remarkable journey from the inside.

Chapter 2
Nuclear Foundations

On October 9, 1941, President Franklin D. Roosevelt changed the course of history by approving a program to develop the world's first atomic bomb. In June 1942, the US Army established the Manhattan Engineering District. US Army General Groves appointed J. Robert Oppenheimer to head up the project's secret weapons laboratory, which became the Los Alamos Laboratory located in New Mexico. Oppenheimer continued to head this monumental effort—known to history as the Manhattan Project—until it reached its goal: to develop a bomb to finally end the war with Japan.

Dr. Frederic de Hoffmann (Fig. 2.1), another Manhattan Project scientist, was studying how to use nuclear energy to produce electrical power. A physicist from Austria, de Hoffmann pursued these goals by founding General Atomic, a research and development facility in La Jolla, California, adjacent to the University of California. Construction of the facility began early in 1950.

At that time, I was enrolled at the University of Wisconsin and was focused on harnessing nuclear energy to produce electrical power. Finding materials that could survive and function at the high temperatures generated within nuclear reactor cores was a major problem.

I spent the summer of 1953 at the Hanford site, a secret section of the Manhattan Project. Operated by the General Electric Company, the Hanford site was located on the south and west sides of the Columbia River in the state of Washington and consisted of large reactors configured to produce plutonium for use in bombs to be stockpiled during the cold war with Russia (Fig. 2.2).

J. Bokros, *Heart of Carbon*, https://doi.org/10.1007/978-3-031-17933-4_2

Fig. 2.1 Frederic de Hoffmann PhD. (Courtesy of General Atomics. All rights reserved)

Grappling with a Serious Problem

I arrived in the spring of 1953, stayed in the barracks in the Tri-Cities, and worked in the 300 Area in the southeast corner of the site. I was to work for Dr. Walter P. Wallace, a mild-mannered metallurgist. It was Dr. Wallace who assigned me to work on what he designated as "a serious problem."

The background to the problem was straightforward enough. The Hanford reactors were assemblies of giant reactors that were separated far from one another. Each reactor contained horizontal tubes, inside of which metallic uranium cylinders, called "slugs," were sealed within aluminum cans. Caps were welded onto the ends of the tubular aluminum cans. Water from the Columbia River served to cool the capped slugs.

The problem manifested itself when the ends of the cans containing the uranium slugs would separate, releasing radioactive material into the Columbia River. Dr. Wallace's description ("a serious problem") was the height of understatement. And now it was a problem that had fallen into my lap.

Galvanized by the challenge, I sectioned the pre-irradiated slugs and analyzed them using metallographic methods. It soon became obvious to me that the anisotropic crystals of uranium formed during heat treatment were developing a pronounced preferential orientation. Their principal axes of rotation were predominantly oriented in the radial direction of the cylindrical slugs, causing them to have anisotropic bulk properties. During thermal cycling inside a reactor, the slugs would expand and contract radially during heating and cooling, respectively. This phenomenon in turn drove an intercrystalline ratcheting mechanism causing the caps to "pop" off.

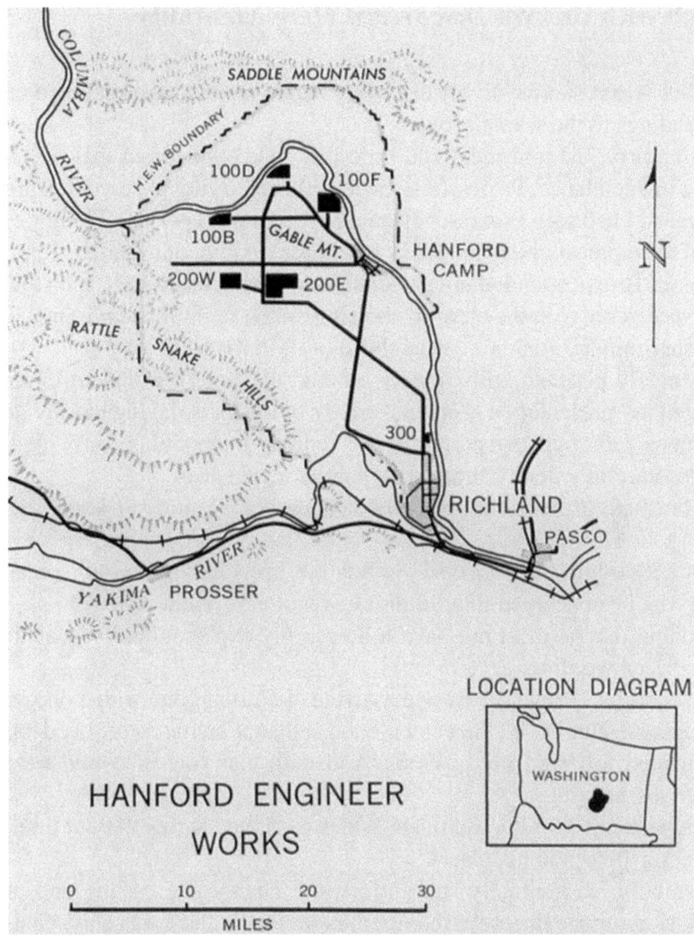

Fig. 2.2 The Hanford Manhattan site. The reactors producing plutonium are located along the Columbia River. I was stationed in the 300 Area located on the southeast corner of the site. (Reproduced with the permission of Kathleen Selbrede)

Using small rods (2 mm in diameter, 20 mm in length) machined from a uranium slug, I measured the thermal expansions in the radial and longitudinal directions using the methodology described by Dr. John Howe et al. at the General Electric Research Laboratory in Schenectady, New York, in an unpublished laboratory report. Upon thermal cycling, the intercrystalline ratcheting mechanism produced the same net dimensional changes as did the Hanford uranium slugs.

On each thermal cycle, internal stresses induced strains that are both elastic and plastic during heating. Upon cooling, the elastic strains were recovered, *but the plastic strains were not*. As cycling continued, minute plastic residual strains accumulated, slowly but inexorably increasing the length of the slug. When the slugs' cumulative strains exceeded the fracture strain of the welds holding the retentive caps in place, the caps broke off.

A Brush with the *Not Discovered Here* Mentality

Dr. Wallace was ecstatic with my discovery and to my dismay scheduled me to present my findings to the whole group.

"Don't worry," he told me, "you'll do fine." He pointed out that one consultant who'd be in attendance, Professor Brown, had been trying to solve this problem for several years. I had seen Professor Brown but had never spoken with him. He always appeared to be preoccupied and fairly indifferent to my contributions to the project.

Professor Brown attended my presentation wearing his tweedy sports jacket with leather patches sewn on the elbows, chewing on the stem of an unlit pipe. I saw him as an absent-minded genius … who chose to sit in the front row!

I was nearly petrified with anxiety but had the benefit of having practiced and memorized my presentation. I stayed strictly to the facts, laying out my analysis of the problem, tethering my points to the tedious methodology provided by John Howe, and carefully demonstrating the failure mechanism.

As I finished, Professor Brown frowned sternly. "You don't know the history," he said.

I admitted to him that indeed I did not, but I privately found this objection perplexing: why be concerned that I didn't know or care about history?

Reasoning that he must not have followed my explanation, I repeated verbatim my memorized wording.

"Preposterous," Professor Brown asserted, dismissing me with a one-word summary judgment. But Dr. Wallace intervened at that point to rescue me, declaring that *he* had indeed followed my analysis. And with that vote of confidence, Wallace adjourned the meeting.

As we walked out of the room, Dr. Wallace grinned at me. "About time someone put Professor Brown in his place!"

The solution was actually straightforward: change the casting and processing variables to eliminate the preferred orientation. Dr. Wallace was quite pleased to put that "serious problem" behind him. I returned to the University of Wisconsin, but I was not to forget him.

Another Rung on the Academic Ladder

After I graduated from the University of Wisconsin with a Bachelor of Science degree, the head of the Department of Mining and Metallurgy called me into his office and inquired about my plans for the future. He was already in his 70s and had acquired a perpetual cough, the price he paid for dedicating so much of his time working in dusty mines.

I was not interested in an advanced degree from Wisconsin. I didn't admit to the department head how thoroughly unimpressed I was with their graduate curriculum. My actual hope was to gain some industrial experience in the nuclear field.

As if he hadn't heard what I said, he asserted that given my academic record, I needed to stay on long enough to at least earn a master's degree.

His approach was father-like. When I pointed out to him that I had no money to continue at the university, he simply picked up the phone and told whoever was on the other end, "I have a young fellow here that needs a stipend." He then hung up and said, "You're all set."

As I left his office, I was thinking sarcastically, "That's great!"

In the fall of 1954, I completed the master's program, developing a method to study the adsorption of carbon atoms at the intercrystallite boundaries in polycrystalline iron, at elevated temperatures, using the carbon-14 radioactive isotope as a tracer. The experimental procedure was published in the *Journal of Metals* in 1956 [1].

More Experience with Reactors

In 1955, I took a position at Atomics International, a Division of North American Aviation, located in Downey, California. Atomics International was developing a power reactor under a contract with the US Atomic Energy Commission. The reactor was moderated by graphite, with an enriched uranium-oxide fuel, and cooled by molten sodium metal. It was called the Sodium Reactor Experiment … and it richly earned that *experimental* epithet.

Molten sodium and graphite are incompatible, so the graphite moderator had to be sealed in zirconium cans. My task was to ascertain whether that solution would work. If the zirconium cans failed, molten sodium metal would intercalate with the graphite, causing an expansion that would jeopardize the viability of the reactor core.

A further contractual obligation under this project was to produce 1000 °F steam, further pushing the environmental envelope we were working with.

I worked on the problem from 1955 to 1958. My evaluation indicated that at 950 °F, the viability of the zirconium can was dubious [2, 3]. I was convinced that at 1000 °F, it was just a matter of time to failure. It was my responsibility to sign-off on the cans before going to full power, and this, I would not do. I was, as they say, "Johnny on the spot," and Johnny would not budge.

As chance would have it, Dr. Wallace, with whom I had worked at the Hanford site, had been selected to join General Atomic in La Jolla, California (Fig. 2.3). He called me in 1958, inviting me to join him there.

Frederic de Hoffmann had put together a consortium of private utilities, along with two oil companies and General Dynamics. Initially, a portion of General Atomic employees was to include colleagues from the Manhattan Project. Dr. de Hoffmann's confident and charismatic personality attracted leaders in science and industry to work with him at General Atomic. Hans Bethe and Enrico Fermi were among the dignitaries enlisted as honorary consultants. It seemed inevitable that General Atomic's John Jay Hopkins Laboratory for Pure and Applied Science would emerge as a "fountain" from which new ideas and technologies would bubble up.

Fig. 2.3 The General Atomic company, La Jolla, California. (Courtesy of General Atomics. All rights reserved)

When I arrived at General Atomic, Dr. Wallace tasked me to evaluate metallic and refractory metallic alloys for high-temperature use in contact with graphite in nuclear reactors [4–6] and to conduct studies of a unique alloy of uranium-zirconium-hydrogen [7–9]. The latter was the fuel material used in General Atomic's water-cooled TRIGA Research Reactor and elsewhere in a wide variety of research endeavors.

The manager of the TRIGA project used the TRIGA reactor to demonstrate the safety of General Atomic's HTGR because it could simulate the safety of that reactor.

The core of the TRIGA reactor was made up of rods of uranium-zirconium-hydrogen fuel immersed in a deep tank of clear water with a circular viewing platform positioned above, so that the core was visible to observers. Because of the unique hydride fuel, the TRIGA core mimicked the core of an HTGR with respect to the "nuclear Doppler effect," so that if the control rods were extracted to produce a runaway situation, the Doppler effect would squelch the nuclear chain reaction and shut the reactor down. I didn't understand the Doppler effect but accepted the premise underlying it.

For the demonstration, the manager would dim the lights and then rapidly extract the control rods, causing a runaway effect that produced an intense blue flash. As the viewer stepped back, the Doppler effect would extinguish the flash, and the lighting would return to normal. This very dramatic effect reminded me of the charismatic personality of Frederic de Hoffmann.

The Final Rung of the Academic Ladder

In 1959 during all this excitement, Professor David Ragone, PhD (Fig. 2.4), a Staff Member from the University of Michigan, approached me. In the course of our conversation, he discovered a fly in the ointment: despite my being a staff member, I did not have a PhD.

"A PhD is like a union card," he asserted, and one that should be required for General Atomic's staff members.

General Atomic was organized along the same lines as the Manhattan Project. There were Staff Members (like full professors) and Staff Associates (like associate professors). Everyone else was simply a technician.

As it happened, I shared an office with a PhD from the University of California at Berkeley. When he heard of Ragone's remark, with no prompting whatsoever, he directed me to get in contact with Earl Parker, a Professor of Materials Science at Berkeley. I contacted Parker and joined the UC Berkeley Doctorate Program in the fall of 1960.

I took a double course load, identified the hurdles, and laid out a plan that would allow me to submit my dissertation to Professor Parker by the fall of 1962.

In 1960, Professor Parker gave me directions to select a thesis subject that would enlighten us as to why "martensite" was so hard. Martensite is a very hard phase in the iron-carbon binary alloy system, important in the iron and steel industry. He was not aware that I had since changed course, nor was he aware that I had changed my thesis objective since he issued his instructions to me. Instead of analyzing hardness, I discovered a different but related phenomenon concerning martensite. In the fall of 1962, with some anxiety, I laid my thesis out on his large conference table.

He looked at my thesis and stared back at me.

"This is not what I told you to do!"

Fig. 2.4 Professor David Ragone PhD. (Reproduced with the permission of David Ragone)

There was a long pause as he examined my elaborate layout. Without looking up, he stated, "There's a mistake here."

After another long pause, he finally said, "I think you can get a thesis out of this."

I did what he told me, and the resulting paper was published in *Acta Met* in 1963 [10]. The topic was an elucidation of the martensite "burst" transformation and *not* why martensite was so hard. Parker would later be quoted as saying that the paper "was the most important advance in the field in recent times."

Diving Back into Industry

Because of my father's illness, I took Professor Parker's suggestion to go to a technical laboratory instead of returning to General Atomic. The lab happened to be near the family home of my wife. By then, we had two young daughters, Kathleen and Carol, and were expecting our third, Linda. Shortly after our move, my father passed.

I called Ragone. "Where have you been?" he asked. I lamented that the laboratory where I worked was not comparable to General Atomic. "Okay," he replied, "but when you come back, I need you to work on soot." I had no idea what kind of "soot" Ragone was talking about.

The reactor envisioned at the time was called the HTGR. It was to operate at very high temperatures (1200–1400 °C) and have a graphite moderator. Uranium-thorium carbide was to be the fuel. The uranium-thorium combination didn't just provide heat. The nuclear reactions produced additional fissionable atoms so that the reactor could breed more fissionable atoms, i.e., it was a "breeder." It was cooled by helium gas. Because of the high temperature, it would represent the most efficient reactor ever developed. It was also envisioned to be the safest.

The fuel, tiny spheres of uranium-thorium carbide, needed to be protected from hydrolyzing during manufacture. If a refractory coating could be developed that was capable of retaining fission fragments so they weren't released to contaminate the reactor system, the reactor core's design would be greatly simplified. The coating for the particles was to be pyrolytic carbon, a material yet to be developed.

Given these parameters, it was natural to explore the use of a high-temperature furnace with a fluid bed system designed to deposit carbon. A vertical tube furnace would contain a fluidized bed of tiny carbide fuel particles (Fig. 2.5). The fluidizing gas would contain a hydrocarbon that, because of the high temperatures, would deposit carbon on the particles that made up the bed (Fig. 2.6). Although rudimentary experimental processes had been tried [11], a new special process would be required.

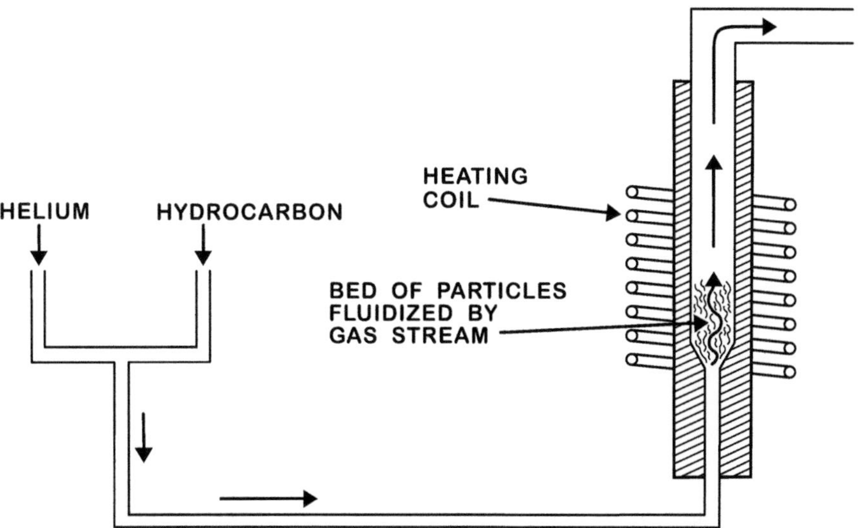

Fig. 2.5 Schematic diagram showing particles being coated with pyrolytic carbon in a fluidized bed. ("Applications of Pyrolytic Carbon in Artificial Heart Valves: A Status Report" is reprinted by courtesy of Carnegie Mellon University Press, Pittsburgh, Pennsylvania)

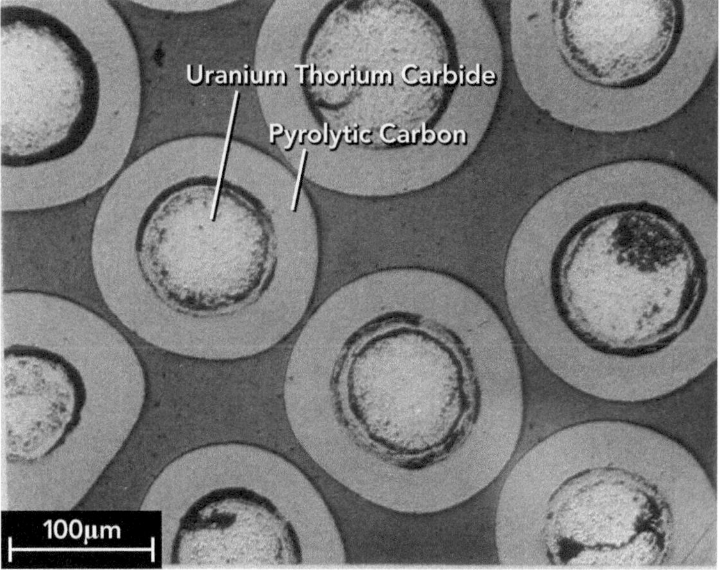

Fig. 2.6 Photomicrograph of a cross-section of spherical uranium-thorium carbide particles coated with pyrolytic carbon. (Reprinted from *Chemical and Mechanical Behavior of Inorganic Materials*, 1969, with the permission of John Wiley & Sons-Books)

References

1. Bokros JC, Rosenthal PC. Liquid emulsion autoradiography of metals. J Metals. 1956;8:286.
2. Bokros JC. Critical recrystallization of zirconium. Trans AIME. 1960;218:351.
3. Bokros JC. Effect of sodium on the mechanical properties of zirconium. Corrosion. 1961;17:131.
4. Bokros JC, Wallace WP. High temperature, high pressure oxidation of alloys caused by carbon–dioxide. Corrosion. 1960;16:117.
5. Bokros JC. Graphite-metal compatibility at high temperatures. J Nucl Mater. 1961;3:89.
6. Meyer RA, Bokros JC. Chapter 15: graphite-metal and graphite-molten salt systems. In: Nightengale RE, editor. Graphite. New York: U.S. Atomic Energy Commission, Academic Press; 1962. p. 445.
7. Bokros JC. Creep properties of uranium-zirconium-hydrogen alloy. J Nucl Mater. 1961;3:216.
8. Bokros JC. Transformation kinetics of a zirconium-uranium-hydrogen alloy. J Nucl Mater. 1961;3:320.
9. Bokros JC, Merten U. Hydrogen migration in zirconium-hydrogen-uranium alloys in thermal gradients. J Nucl Mater. 1963;10:201.
10. Bokros JC, Parker ER. Mechanism of the martensite burst transformation in Fe-Ni single crystals. Acta Metall. 1963;11:1291.
11. Oxley JH, Secrest AC, Veigel ND, Blocher JM. J Am Inst Chem Eng. 1961;7(3):498.

Chapter 3
Origin and Characterization of Medical Carbons

By the time I returned to General Atomic in early 1963, Dr. David Ragone was Head of Metallurgy and had assumed overall responsibility for the fuel development group. That group was headed by a staff member who held a baccalaureate degree in chemistry.

Dr. Ragone assigned me to fuel development, where I'd be reporting to a staff member at the administrative level, but would be acting independently to get a handle on the science of "soot." The arrangement was strikingly parallel to my work under Professor Parker at Berkeley, who had tasked me to discover "what makes martensite so hard."

I was to learn about soot from a famous woman scientist, Rosalind E. Franklin. She was best known for her work on the x-ray diffraction images of DNA, which led to the discovery of the double helix, for which James Watson, Francis Crick, and Maurice Wilkens shared a Nobel Prize in 1962.

The omission of Franklin's contributions did not go unnoticed. The Nobel Prize was given in 1962. Traditionally, Nobel laureates were not made posthumously. Watson suggested that Franklin would have ideally been awarded a Nobel Prize in Chemistry along with Wilkens. But Franklin was taken by ovarian cancer in April of 1958 (age of 37) attributable to x-ray radiation experienced in her research endeavors.

After her work on DNA, Franklin pioneered work on the molecular structure of viruses. Aaron Klug continued her work in that regard, winning the Nobel Prize in Chemistry in 1982.

There were allegations of sexism perhaps due to Franklin's elevated respect among her male peers. Not so incidentally, she resented being called Rosy in lieu of Rosalind (https://en.wikipedia.org/wiki/rosalind.franklin).

Franklin's term for soot might have been "quasi-crystalline pyrolytic carbon deposited at high temperatures by the pyrolysis of a gaseous hydrocarbon at high temperatures in a fluidized bed." As she pursued a PhD thesis, Franklin's challenge was to understand the structure of coal, published as "The interpretation of diffuse

© The Author(s), under exclusive license to Springer Nature Switzerland AG 2023

J. Bokros, *Heart of Carbon*, https://doi.org/10.1007/978-3-031-17933-4_3

X-ray diagrams of carbon" in *Acta Crystallographica* in 1950 [1]. This paper, along with three others [2–4], provided a sound basis for our studies on soot. She and I were on the same page, so I followed her lead [5–7].

Ragone recognized that a carbon coating for the fuel might retain fission products and thus prevent the contamination of the reactor system. At this developmental stage, carbon coating was only being used to protect the uranium-thorium carbide from hydrolyzing during processing.

In 1963, as directed by Ragone, I undertook a study of the structure and properties of carbon that was produced using fluid bed technology. Carbons formed in fluidized beds have a continuum of complex structures, each intimately related to the mechanism involved in the deposition.

Most of the atoms are arranged in planar hexagonal arrays, linked together by strong covalent bonds (Fig. 3.1). Some arrays may be wrinkled and contain vacant lattice sites. Most of them, called crystallites, are parallel to one another, bound together by weak van der Waals forces. The atoms in one layer are not oriented in any particular way with respect to atoms in adjacent layers. Thus, the crystallites are deemed to be "turbostratic." In many deposits, some of the layers are single layers as well as amorphous carbon. A schematic of an aggregate is shown on the lower right in Fig. 3.1d.

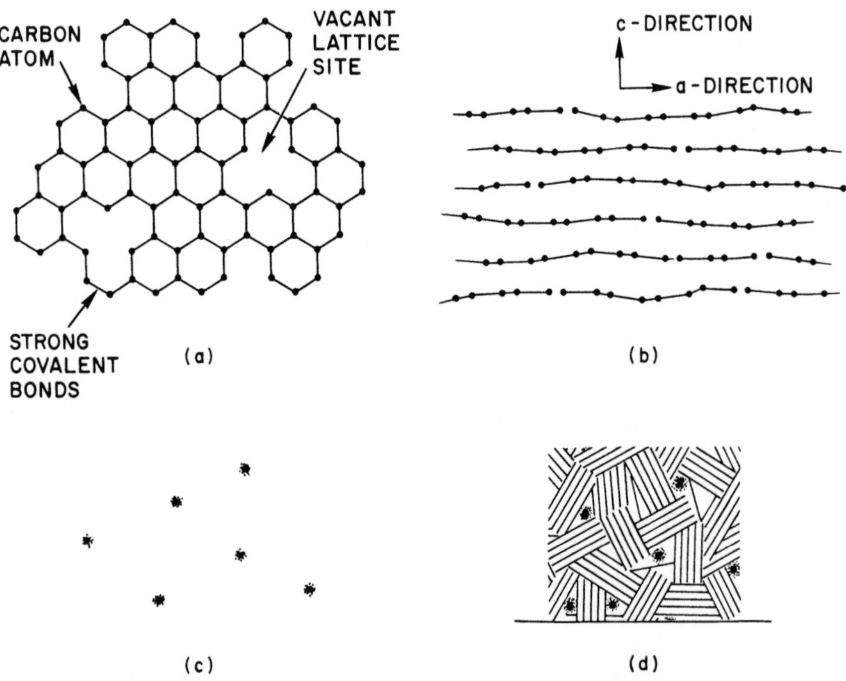

Fig. 3.1 Diagrams of possible atomic arrangements in pyrolytic carbon (**a**) a single layer plane, (**b**) parallel layers in a crystallite, (**c**) amorphous carbon, and (**d**) an aggregate of crystallite's single layers and amorphous carbon. (Reprinted from *Chemistry and Physics of Carbon: A Series of Advances*, 9, 1973, with the permission of Taylor & Francis Group LLC-Books)

Because of the strong covalent bonding *within* the layers and the weak van der Waals bonding *between* them, the properties of the individual crystallites are anisotropic. Therefore, the properties of an aggregate vary depending on the degree of preferred orientation, as well as upon the presence of single layer and amorphous carbon.

Irradiation with fast neutrons causes the crystallites to expand perpendicular to the layer planes and contract in the parallel direction. I was quite surprised when a disc (with a high degree of preferred orientation) was irradiated at high temperature with a high dose of fast neutrons: it shrunk diametrically and grew in thickness. The disc became a rod! And despite the considerable deformation, there was no evidence of internal fracturing.

Quasi-crystalline carbon is truly remarkable. It can be manipulated through processing to get anything you'd like—perhaps even diamond.

Black Box Analysis

As I reviewed the literature, I was amazed at the far-ranging variations in the structure and properties of such carbons. The complexity of the pyrolysis mechanisms was equally staggering. It was too complex for modeling, so I decided to set up the process as a "black box."

Tech Note: Those interested can refer to *Section II, Mechanisms of Carbon Formation from Gaseous Hydrocarbons* (in *Chemistry and Physics of Carbon*, Vol 5, pages 3–7) [8].

The approach requires the identification of the independent variables that control the reactions in the "black box." The literature indicates that the four independent variables are the temperature of the fluid bed, the concentration of the hydrocarbon that is the carbon source, the total flow rate which determines the contact time in the bed, and the reacting bed surface area (Fig. 3.2).

The first three variables are not a problem because instrumentation for measuring them was available. But the bed surface area inherently increases as carbon is deposited on the bed. To mitigate the effects of the changing surface area, we used as short a deposition time as possible so that the change in surface area during deposition was minimized.

The density of a deposit was determined by a simple sink-float method. The apparent crystallite size and preferred orientation were determined by x-ray diffraction (Fig. 3.3). Tedious x-ray scattering measurements, *à la* Franklin, provided a measure of the amount of disorganized carbon and the number of single layers that were present in the deposits [1–4].

The properties of carbons formed by deposition in a fluidized bed and the techniques used to measure them have been published [8].

The pole figure reflects an important parameter obtained by x-ray diffraction that characterizes a carbon deposit (Fig. 3.4). In an aggregate of crystallites, the pole figure relates the relative number of crystallite "poles" (the normals perpendicular

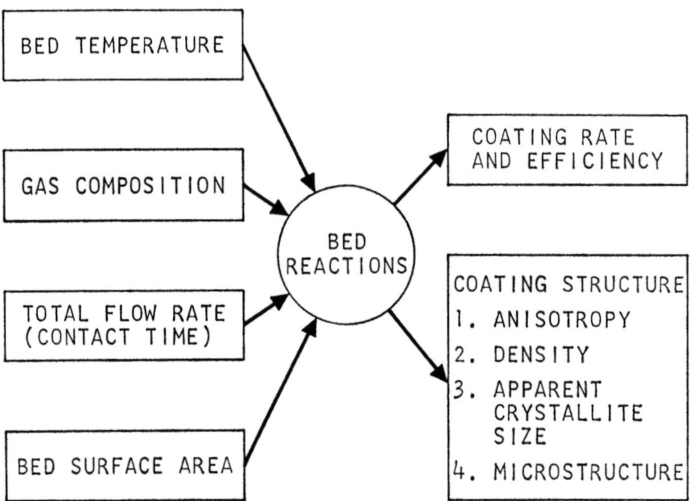

Fig. 3.2 Diagram relating the four coating variables, the bed reactions, the rate of deposition, and the structure of the carbon deposited. (Reprinted from *Chemical and Mechanical Behavior of Inorganic Materials*, 1969, with the permission of John Wiley & Sons-Books)

to the layer planes in a crystallite) inclined to the z-axis, at the angle Φ. Those unfamiliar with the subject should study and understand this concept because the Bacon Anisotropy Factor, BAF, which is derived from the pole figure, relates all the properties of an aggregate to the properties of its individual crystallites [8].

In the "black box" approach, the structure and properties of the deposited carbon were measured as a function of the independent variables, so that three-dimensional plots that related each of the important properties to their deposition origin could be generated (Fig. 3.5) [8].

The literature suggested that the pyrolysis of gaseous hydrocarbon deposits carbon on surfaces in the chamber that are energetically favored and the layers formed in the gas tend to orient parallel to the surface on which they are deposited. Accordingly, most deposits have a preferred orientation of their layers parallel to the surface on which they are deposited. As a result, they are not suited for use in the core of a nuclear reactor that is exposed to high doses of fast (high-energy) neutron radiation.

In 1963, I found the evidence of a truly isotropic pyrolytic carbon (the horizontal line at the top of Fig. 3.4) [9]. In this case, the energetic favorability to deposit directly on a surface was overcome by supersaturating the gas with carbon radicals such that the deposit is made up of droplets formed in the gas phase and then deposited on the surfaces in the bed [7].

To some, these phenomenological studies were viewed as "black magic." I was blessed to be labeled "Black Jack, the Carbon Quack." I actually smiled when called a quack and took it as a wry compliment.

When I showed the x-ray diffraction pattern of this isotropic carbon to the leader of the fuel development project, he declared "Bokros, you raise more questions than you answer!" As Ragone had suggested, he really did need help.

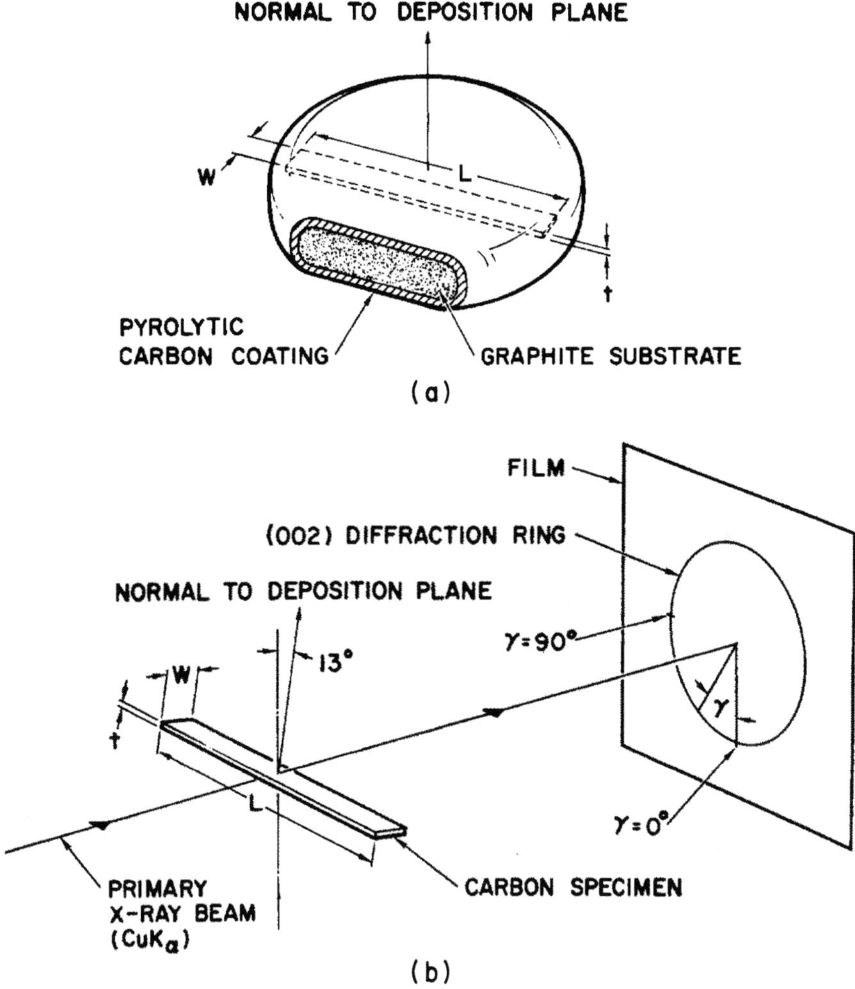

Fig. 3.3 Diagram showing (**a**) carbon-coated graphite disc and the position from which the x-ray specimen is cut from the coating and (**b**) experimental arrangements of a carbon specimen for preferred orientation and apparent crystallite size measurements using CuK_α radiation. (Reprinted from *Carbon*. Absorption factors for a modified bacon preferred-orientation technique, JC Bokros, Oct 1, 1965; 3(2):167–174 with permission from Elsevier)

New Directions Based on New Evidence

But the reality was that isotropic carbons are more dimensionally stable at high temperatures and high levels of neutron irradiation than those that are anisotropic. This finding changed the direction of General Atomic's fuel development effort.

The new objective was to develop an isotropic coating to contain fission products that would remain intact during years of operation at high temperatures, all while being subjected to intense irradiation by neutrons. More details are available in Ref. [10].

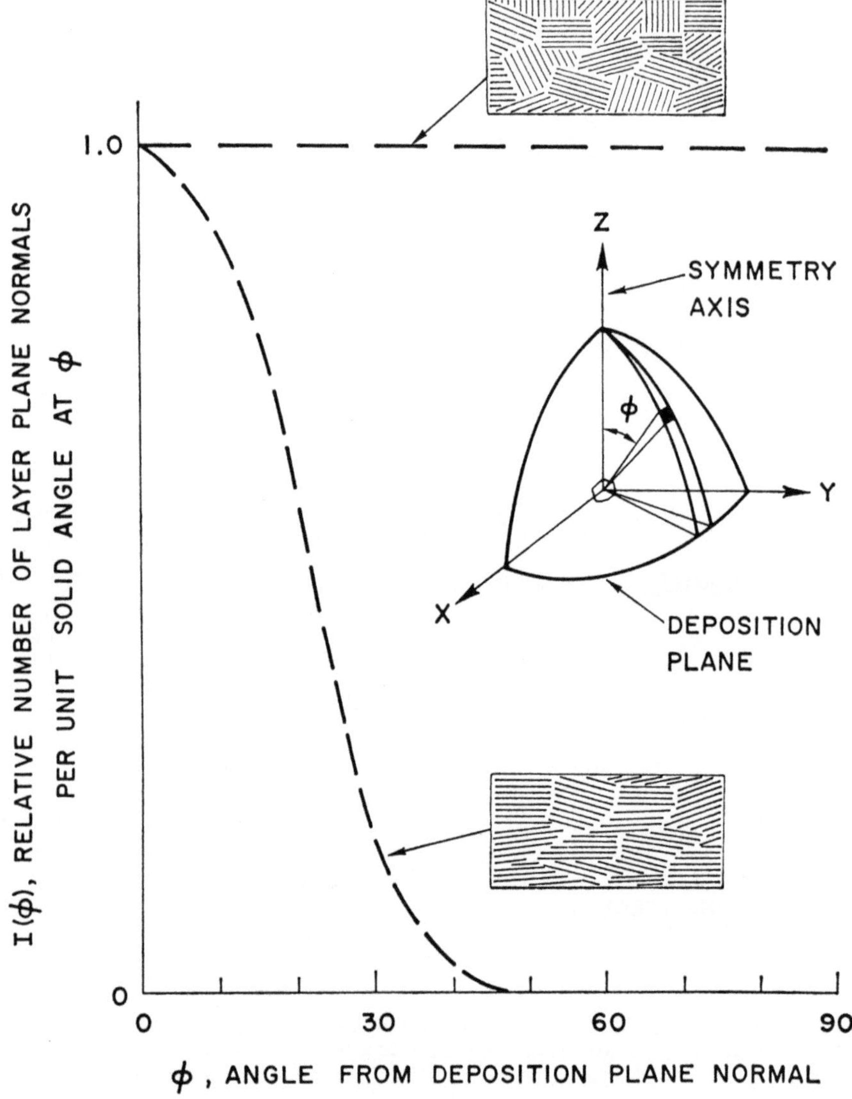

Fig. 3.4 Pole figure for an isotropic carbon and a carbon with considerable preferred orientation (inset shows relationship between deposition plane and angle Φ in an aggregate of carbon crystallites). (Reprinted from the *Journal of Biomedical Materials Research* Part A. Correlations between blood compatibility and heparin adsorptivity for an impermeable isotropic pyrolytic carbon, MD Ramos, KD Vos, AM Fadall, et al., Sep 13, 2004;3(3) with permission from John Wiley & Sons)

By characterizing these near-isotropic deposits, it was possible to calculate an aggregate's dimensional changes (induced by irradiation by fast high-energy neutrons at high temperature) from previously determined data relating the individual crystallite shape-change data to the fast neutron irradiation dose.

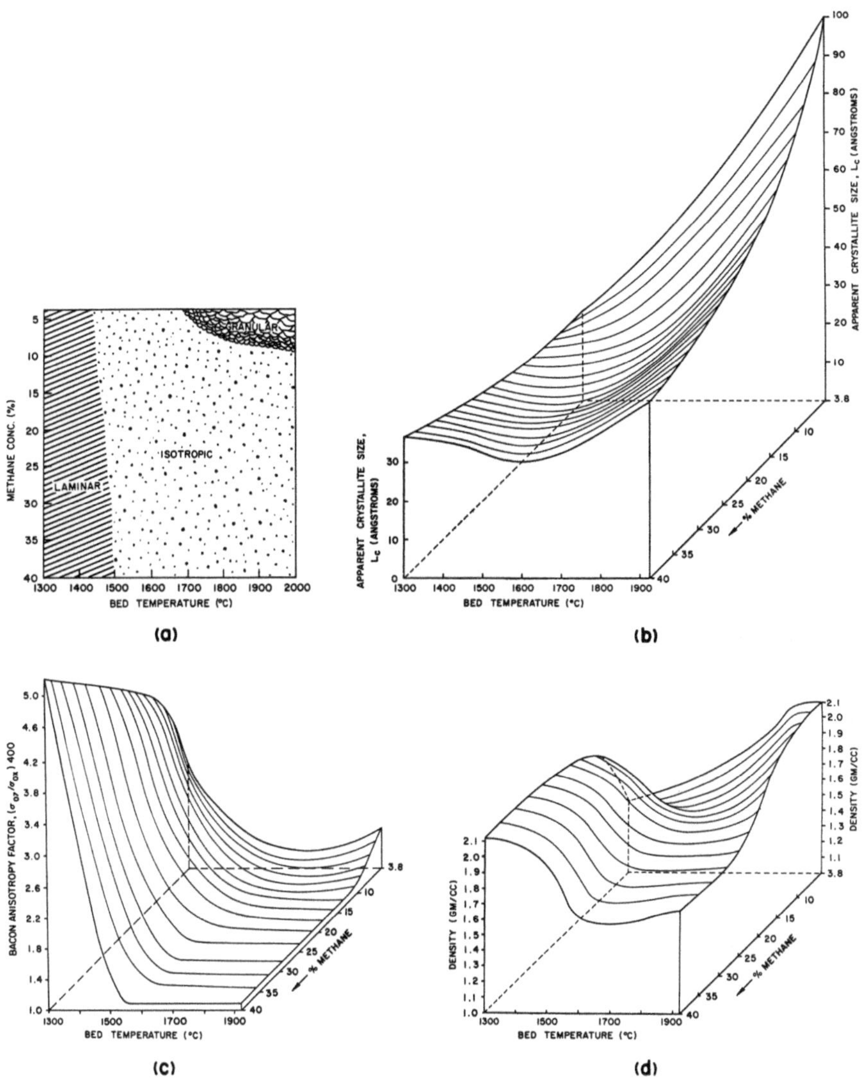

Fig. 3.5 Diagrams showing the relationship between the structure of the carbon and its deposition conditions: (**a**) microstructure, (**b**) apparent crystallite size, (**c**) anisotropy factor, and (**d**) density. (Reprinted from *Chemical and Mechanical Behavior of Inorganic Materials*, 1969, with the permission of John Wiley & Sons-Books)

A demonstration for those interested is in Fig. 3.6 as a plot of the dimensional changes versus preferred orientation parameter, R_x. The plot is Fig. 5 from Ref. [10]. For the perfect isotropic case, $R_x = 2/3$, and for dense carbons, the dimensional changes are small. From Fig. 3.6, note that when $R_x = 2/3$ and the deposits are sufficiently dense, the deformation (theoretically at least) equals zero.

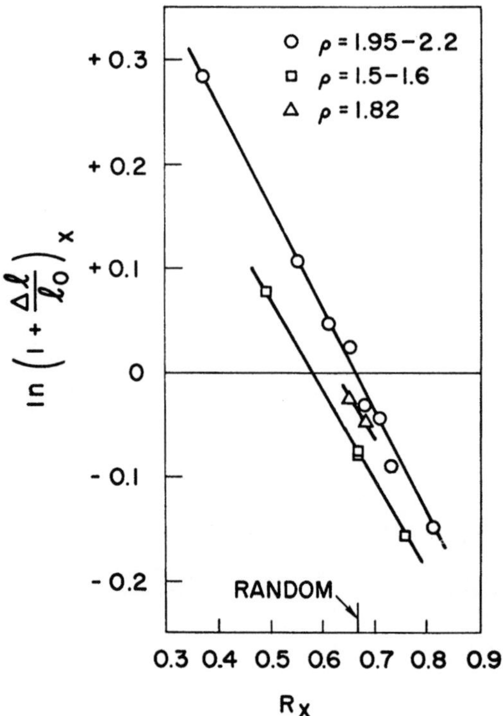

Fig. 3.6 Plot of dimensional changes, $\ln[1 + (\Delta l/l_o)]_x$, versus preferred orientation parameter, R_x, for as-deposited carbons with varying density irradiated at 1040 °C to a neutron exposure of 2.4×10^{21} nvt ($E > 0.18$ MeV). (Reprinted from *Carbon*. Radiation-induced dimensional changes in pyrolytic carbons deposited in a fluidized bed, JC Bokros, RJ Price, Oct 1, 1966;4(3):441–454 with permission from Elsevier)

References

1. Franklin RE. The interpretation of diffuse x-ray diagrams of carbon. Acta Cryst. 1950;3(2):107–21.
2. Franklin RE. On the structure of carbon. J Chim Phys Phys-Chim Biol. 1950;47(5,6):573–5.
3. Franklin RE. A rapid approximate method for correcting the low-angle scattering measurements for the influence of the finite height of the x-ray beam. Acta Cryst. 1950;3(2):158–9.
4. Franklin RE. Influence of the bonding electrons on the scattering of x-rays by carbon. Nature. 1950;165(4185):71–2.
5. Bokros JC. The structure of pyrolytic carbon deposited in a fluidized bed. Carbon. 1965;3:17.
6. Bokros JC. Absorption factors for a modified bacon preferred-orientation technique. Carbon. 1965;3:167.
7. Bokros JC. Variations in the crystallinity of carbons deposited in fluidized beds. Carbon. 1965;3:201.
8. Bokros JC. Chapter 1: deposition, structure, and properties of pyrolytic carbon. In: Walker PL, editor. Chemistry and physics of carbon, vol. 5. New York: Marcel Dekker, Inc; 1969.
9. Bokros JC. Random pyrolytic carbon. Nature. 1964;202:1004.
10. Goeddel WV, Bokros JC. AIME nuclear metallurgy symposium. Delavan, Wisconsin, October 1966. In: Holden AN, editor. High temperature nuclear fuels, vol. 42. New York: Gordon and Breach; 1966. p. 85–104.

Chapter 4
Gott's Early Experience: A Lesson in Serendipity

I became aware of Vincent L. Gott MD from an abstract submitted to the Carbon Journal in 1963 for a paper to be presented at a Carbon Conference to be held in Pittsburgh, Pennsylvania, in June of that year. The abstract caught the eye of Bill Ellis, a co-worker, who passed it on to me.

The abstract from Vincent Gott and Ronald Daggett, Professor of Plastic Engineering at the University of Wisconsin, described the development of a graphite-pigmented paint used to coat a polycarbonate housing for a heart valve replacement.

I sent an inquiry addressed to Dr. Gott at the University of Wisconsin, describing the carbon coating we were developing for nuclear fuel. Intrigued, he asked for short tubes to test in canine venae cavae. We didn't really understand the test but nonetheless fabricated the requested tubes (Fig. 4.1) and sent them to Dr. Gott for evaluation.

More than a year passed and I had not heard from Dr. Gott. I called the University and discovered that Dr. Gott had taken a position at the Johns Hopkins Medical Institutes located in Baltimore, Maryland. When I finally reached Dr. Gott (Fig. 4.2), he was apologetic about the long delay, but he was quite animated in explaining that our materials were "just amazing." In fact, he wanted to know if I could visit him in Baltimore.

The excitement in his voice got *me* excited, so I flew out to Baltimore.

The Artificial Heart Program

The Johns Hopkins facility was impressive, as was Gott's title: Surgeon-in-Charge. His office was large and was dominated by a huge conference table. One wall was lined with files. After touring his facility, we returned to his office, where he showed me a black heart valve prosthesis (Fig. 4.3) he had been developing. However, he

J. Bokros, *Heart of Carbon*, https://doi.org/10.1007/978-3-031-17933-4_4

Fig. 4.1 Tubes sent to Vincent L Gott MD for test and evaluation. (Used with the permission of Jack Bokros)

Fig. 4.2 Vincent L Gott MD. (Used with the permission of Vincent Gott, MD)

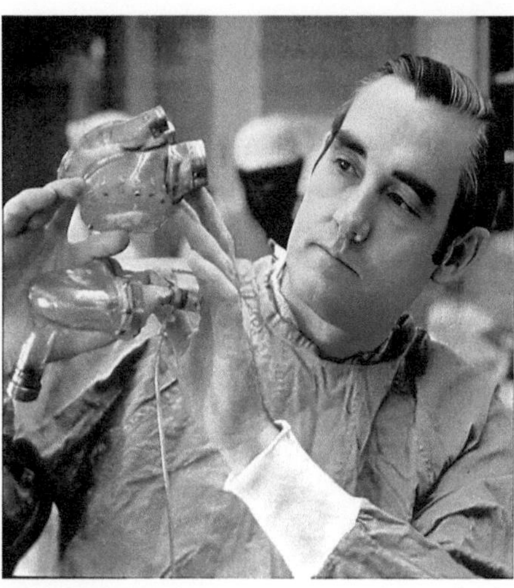

Fig. 4.3 The Gott bileaflet valve was introduced in the early 1960s. The black housing is polycarbonate coated with a carbon pigmented paint. The leaflets are silicone rubber with a fabric reinforcement. Clinically, the valve was dipped in benzalkonium chloride and then in heparin, which provided a thromboresistant coating called GBH developed by Gott and colleagues. (Used with the permission of Vincent Gott, MD)

was much more interested in General Atomic and hearing about the carbon coating that was being developed.

As I was finishing, he impatiently put up his hand and turned to his secretary, asking her to "connect me with Frank." He meant Frank W. Hastings, MD, of the NIH and Chief of the Artificial Heart Program.

"I have Dr. Bokros here from General Atomic whom I've told you about." Dr. Gott said. "I'm very impressed with their carbon material."

Then after a long pause, Gott recommended to Hastings that General Atomic be contracted to supply the Artificial Heart Program with their material evaluations.

I returned to San Diego with a very good feeling, as if we were embarking on a new adventure.

While waiting for a response from NIH, I did a quick study of Gott's background. I learned that after receiving an MD from Yale, he had become part of an innovative group of doctors who had trained with Dr. C. Walton Lillehei, considered to be the father of open heart surgery. As an intern, Gott had observed Lillehei perform the first open heart surgery in 1954. Afterward, Lillehei was impressed with Gott's anatomic illustrations of the cross-circulation procedure he had used to correct a ventricular septal defect in an 11-year-old boy. The boy's father served as the oxygenator. This was a "landmark," contributing to Lillehei's increasing fame. That wouldn't be the last we would hear of "Walt" Lillehei.

The Gordon Conference

My first taste of what I was to encounter on the Artificial Heart Program came when Dr. Gott invited me to attend an informal Gordon Conference to be held at a small college in Vermont.

The name of the college escapes me, but I have a vivid memory of the housing that was provided. The small dormitory room was equipped with a cot that had a thin mattress supported by a single layer of springs. Its intricate network of connecting wires provided a hammock-like support, but it was plagued by a squeaky response to every movement. Perhaps college administrators believed that Spartan accommodations would keep students better focused on their lessons.

As I was a first-time visitor to this event, I sat in the back of the auditorium to observe and be inconspicuous.

The Artificial Heart Program Principal Investigator opened the conference with an elaborate, detailed outline of the plan that the Artificial Heart Program was to follow. The tasks and responsibilities for each of the centers of excellence were beautifully depicted in a rapid sequence of colorful slides. I was stunned by the project's complexity, but was relieved when the Principal Investigator put the audience at ease by declaring that the Gordon Conference was very informal. The presentations could be free-wheeling, as no published proceedings would be forthcoming, and the publishing suitability of new, unpublished materials would not be jeopardized. In fact, he invited "brainstorming" to encourage an open dialogue to challenge presenters and added that even the timing of the sessions was to be flexible.

At the break, Vincent Gott sought me out, pointing out that I looked lonesome. He urged me to move up to the front of the hall to better hear his upcoming opening remarks.

Gott provided the lead. The major problem faced by investigators was finding materials that were blood compatible and useful in the construction of an implantable total artificial heart. His GBH material was showing promise, so much of the discussion focused on how the heparin was bound in the coating. Was it bound covalently, or was it possibly bound by an ionic bond?

A polymer under development that sought to mimic GBH using a urethane, was described and proposed as a possible solution route. The ensuing discussions, coming from an audience across a wide range of disciplines who were urged to get involved, were often off-point and thus not productive. As I saw how this was playing out, my initial anxiety (about General Atomic's work being insufficiently relevant to discuss here at a *medical* conference) was reversed, giving me confidence that our carbon technologies could indeed meet the challenging criteria set forth by the Artificial Heart Program.

The next speaker was Philip Sawyer, a surgeon from the Downstate Medical Center in New York.

His opening statement: "As I was browsing through the *Handbook of Chemistry and Physics*, I found a table that listed the electromotive series of metals. I surmised that the clotting of a metal might be related to its electronegativity." He quoted a paper by Abramson, published in a 1925 issue of *Experimental Medicine*, which developed a theory of charge characteristics that explained blood clotting.

I didn't follow Sawyer's idea, but chuckled when a voice from the back of the auditorium admonished the speaker in a jocular yet critical tone: "Phil, I think you're on the wrong page of the book."

The critic was challenging the notion that clotting caused by a metal might depend on the order the metal appeared in the electromotive series. Such a comment emanating from the audience was legitimate within the explicitly informal setting of a Gordon Conference, but the destabilizing impact on Dr. Sawyer's presentation was evident; he never fully regained his stride thereafter.

At the intermission, I learned that the voice that had commented on Sawyer's presentation belonged to Leo Vroman.

I approached Dr. Vroman at the next session break. In our conversation, I discovered that he was an accomplished hematologist who had produced a PhD thesis on "Surface contact and thromboplastin formation." (Much later, in collaboration with Edward Leonard, he would study "The behavior of blood and its components at interfaces" and "Blood in contact with natural and artificial surfaces," both to be published in the *Annals of the New York Academy of Sciences*, Vol. 283 in 1977 and Vol. 516 in 1987.)

Vroman, sensing that I felt somewhat out of place, was friendly and quickly became fascinated with General Atomic's new carbon technology. We talked throughout the rest of the conference. I learned of his talents as a poet and discovered that his criticisms were often poetic, whether they were complimentary in tone or not.

I learned later that his first published poem appeared in 1946 and that he was subsequently awarded many Dutch literary prizes for poetry.

In the evening, Vroman reviewed his portfolio of hematological stained slides with me. They were beautifully colored. He asked me to select two. I had them enlarged; they have hung on my office wall(s) for 50 years before age took its toll and their stunning colorations faded. They are now stored in a box.

Another memorable presentation at the Gordon Conference was delivered by C. William (Bill) Hall. He arrived later, flying his own plane in from Houston, where he was the director of the Artificial Heart Program at Baylor College of Medicine with Michael DeBakey.

Hall appeared at the lectern, complete with his western hat and boots, to describe a velour material developed as a tissue-bonding interface for vascular implants. Vroman, to no one's surprise, was poetically critical of Hall's approach but conceded that "it might just work." Hall and Sam Hulbert were to later found the Society for Biomaterials. Bill Hall, Leo Vroman, and Vincent Gott made the Gordon Conference a very rewarding experience.

After returning from Vermont, I located a number of references to Gott's publications [1–4]. One showed how short tubes were implanted in canine venae cavae to measure clotting times. Dr. Gott's experience thus far was that all materials tested had clotted solidly in 2 h. However, using the graphite pigmented paint on tubes that had been sterilized by dipping sequentially in benzalkonium solution and then in heparin solution gave better results. Rather than clotting in 2 h, the tubes stayed free of thrombus for 2 weeks (Fig. 4.4).

The carbon coating was called GBH, with G designating graphite used as a pigment. The graphite pigment made the coating conductive, so that by using a battery wire, an electrical potential could be applied to a coated implant to induce a negative

MATERIAL	TIME OF IMPLANT	AMOUNT OF THROMBUS IN LUMEN
POLYCARBONATE	2 HRS.	● ● ● ● ●
POLYPROPYLENE	2 HRS.	● ● ● ● ●
TEFLON	2 HRS.	● ● ● ● ●
SILICONE RUBBER	2 HRS.	● ● ● ● ●
# 304 STAINLESS STEEL	2 HRS.	● ● ● ● ●
G.B.H. COATED POLYCARBONATE	2 WEEKS	○ ○ ○ ○ ○

Fig. 4.4 In vivo results for rings coated with Gott's GBH coating compared with other commonly used materials tested in canine venae cavae (Gott and Whiffen). (Reprinted from *Chemistry and Physics of Carbon: A Series of Advances*, 9, 1973, with the permission of Taylor & Francis Group LLC-Books)

charge, thought to prevent clotting. It was reported further that, when the charging wire broke, the GBH coating was discovered to be thromboresistant even without the negative charge. It was surmised that the "G" in GBH was a contributor to the thromboresistance in the absence of the charge.

Another report described an experience with a GBH-coated canine pulmonary valve and pulmonary artery prostheses in a dog.

The Gott valve was first used clinically in humans in April of 1963, at the University of Wisconsin.

General Atomic Enters the World of the Artificial Heart Program

In mid-1965, NIH attorneys traveled to General Atomic to discuss a contractual arrangement. At their meeting, General Atomic attorneys made it clear that General Atomic coveted its technology and would accept no contract that jeopardized its proprietary rights. NIH accepted those terms, and a contract was awarded late in 1965. Our association with the Artificial Heart Program [5] turned out to be an adventure.

A colleague and I decided to file an application for patents on the use of General Atomic's pyrolytic carbon coatings for implantable medical devices [6].

This decision turned out to be important.

At about the same time the aforementioned patent was filed, another patent application had been filed by De Laszlo [7], covering a large number of carbons, including pyrolytic carbon. There was, in fact, a conflict between the Bokros/Ellis and the De Laszlo patent applications.

That crucial dispute was resolved in our favor on the basis of a memo that I had written to David Ragone, Head of Metallurgy and Fuel Development, suggesting that General Atomic's carbons and other ceramic materials held promise as biomaterials. The resolution of the conflict hinged on the date of conception, which turned out to be in our favor by about a week in 1965.

Lucky draw, I thought. Had the decision gone the other way, all of the carbon prostheses that were developed over the last five decades might never have been created.

Continuing forward in this vein, we took meticulous pains (at Ragone's direction) to patent each and every innovation, i.e., to develop a blanket of protection around our intellectual property.

In the early 1970s, we obtained broad coverage for the use of Pyrolite Carbon in biomedical prosthetic devices [6, 8–11], with three more patents being issued through 1979 [12–15]. By agreement with NIH, all these patents were owned by General Atomic.

The objective of the program sponsored by the Artificial Heart Program was to determine the relationships between the crystal structure, surface topography, and

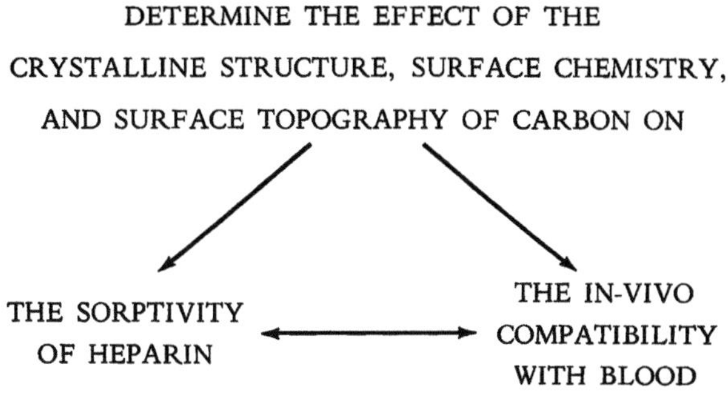

DETERMINE THE EFFECT OF THE

CRYSTALLINE STRUCTURE, SURFACE CHEMISTRY,

AND SURFACE TOPOGRAPHY OF CARBON ON

THE SORPTIVITY   THE IN-VIVO
OF HEPARIN   ← →   COMPATIBILITY
                        WITH BLOOD

Fig. 4.5 Schematic diagram of program research objectives. (Reprinted from *Chemistry and Physics of Carbon: A Series of Advances*, 9, 1973, with the permission of Taylor & Francis Group LLC-Books)

surface chemistry of a variety of different carbonaceous materials and their compatibility with blood (Fig. 4.5).

The assumption originally driving the program plan was that the G in GBH (the graphitic carbon that had been used in the graphite pigmented coating) was able to produce a bond with heparin. To eliminate the reader's confusion upfront, you should understand that this assumption was incorrect. The carbon itself had nothing to do with the bonding of heparin. It was actually the porosity of the coating, not the graphite pigment, which was the source of the thromboresistance.

This reveals the significance of our chapter subtitle, "A Lesson in Serendipity." Science often advances when important things are discovered by accident. The story unfolds step by step as follows.

Following the Evidence Where It Leads

The relevance of the sorptivity of heparin was investigated using radioactive tracer techniques; the retentivity of sorbed heparin was determined using a variety of elutriation methods.

Although a wide variety of carbonaceous materials were investigated, emphasis was given to a family of isotropic carbons developed for coating nuclear fuel particles. These coatings, in their pure form, have fracture strengths in excess of 50,000 psi. Their ability to sustain elastic strains ranging up to 2% made them much tougher than glassy or vitreous carbon or ceramics, which are alternatives that might be considered for use in an artificial heart. Further, the extremely low wear rate of carbon bearing on carbon provides valves with functional lifespans longer than human life expectancies. Isotropic carbon is unique in that its strength is not degraded by cyclic loading—it is immune to fatigue [16–18].

Early results from the Artificial Heart Program study were documented in three publications [19–21], which included discussion of the crystal structure and surface topography of carbon and the methods used in their characterization.

Results for smooth isotropic carbons deposited at relatively low temperature near 1300 °C and cleaned by outgassing are shown in Fig. 4.6. Polished, clean LTI carbons are thromboresistant without further treatment. (LTI carbon is an abbreviation for isotropic carbon deposited at low temperatures, near 1300 °C.) These carbons are impermeable. Heparin is adsorbed on their smooth surface as a monolayer. The adsorption is not crystallographically selective, and adsorbed heparin on the surface did not improve thromboresistance. Neither the heparin adsorption nor the in vivo thromboresistance of LTI carbon is enhanced by the presence of benzalkonium chloride.

Results for smooth LTI carbon surfaces contaminated with oxygen are summarized in Fig. 4.7. The data indicate that chemisorbed oxygen (hydrophilic sites) on LTI carbon surfaces reduces their thromboresistance. Chemisorbed oxygen does not influence the amount of heparin that can be adsorbed on LTI carbon surfaces, and adsorbed heparin on oxidized surfaces does not improve thromboresistance. In fact,

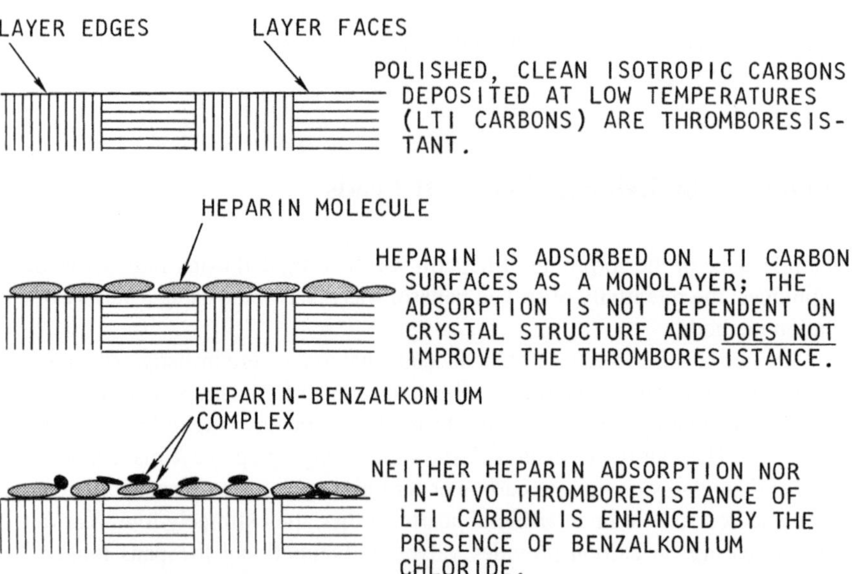

Fig. 4.6 Results for clean smooth LTI* carbon surfaces (*LTI carbon is an abbreviation for isotropic carbon deposited at low temperatures near 1300 °C). (Reprinted from *Chemistry and Physics of Carbon: A Series of Advances*, 9, 1973, with the permission of Taylor & Francis Group LLC-Books)

RESULTS FOR SMOOTH LTI CARBON SURFACES CONTAMINATED WITH OXYGEN

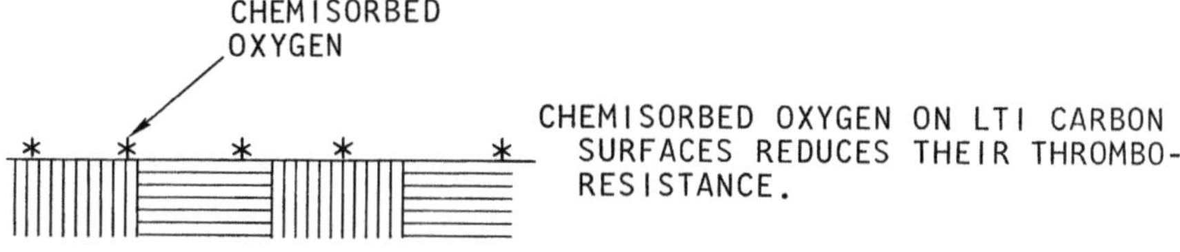

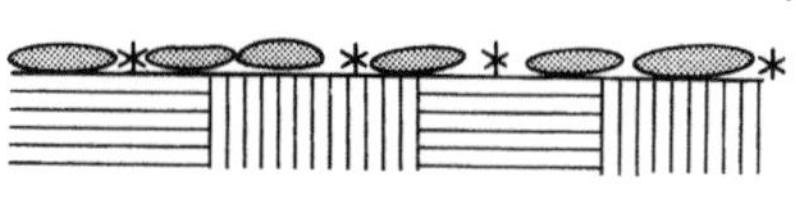

Fig. 4.7 Results for smooth LTI carbon surfaces contaminated with oxygen. (Reprinted from *Chemistry and Physics of Carbon: A Series of Advances*, 9, 1973, with the permission of Taylor & Francis Group LLC-Books)

the "activation sites" that were capable of covalently bonding heparin, if left unbonded, were actually thrombogenic.

Results obtained from studies of rough and porous carbon surfaces are displayed in Fig. 4.8. Rough LTI carbon surfaces are less thromboresistant than polished surfaces, and adsorbed heparin does not improve the thromboresistance of unpolished surfaces.

However, heparin sorption on surfaces with deep, accessible porosity was greatly increased by a surface pretreatment with benzalkonium chloride. This increase was due to the precipitation of a quaternary ammonium salt of heparin in the pores that had a low solubility in water. Rough carbon surfaces are only thromboresistant when the heparin-benzalkonium complex is sorbed in the porosity.

We concluded that surfaces with deep, accessible porosity can be rendered thromboresistant by the GBH treatment. The carbon pigment was not, in itself, an essential factor in achieving thromboresistance.

A large fraction of the heparin sorbed on all the carbon surfaces studied was elutriated by brief washing in plasma; however, a small residue persists for substantially long times. This long-time persistence suggested that some heparin on the surface is sealed in place by a layer of material adsorbed from the plasma.

RESULTS FOR ROUGH AND POROUS
CARBON SURFACES

SURFACE ROUGHNESS DECREASES THE THROMBORESISTANCE OF LTI CARBON SURFACES; ADSORBED HEPARIN (OR HEPARIN-BENZALKONIUM COMPLEX) DOES NOT INCREASE THEIR THROMBO-RESISTANCE.

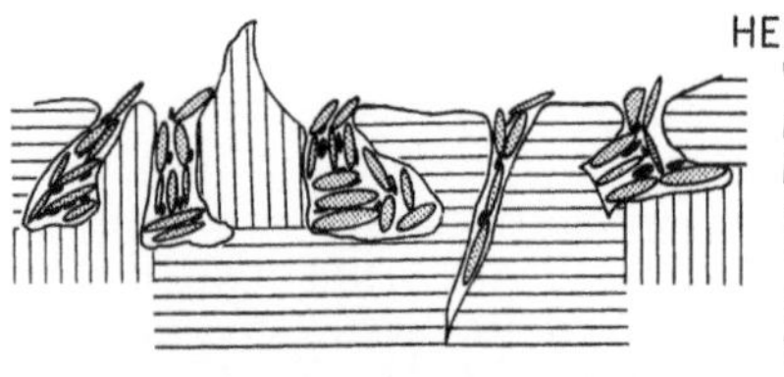

HEPARIN SORPTION ON CARBON SURFACES WITH DEEP ACCESSIBLE POROSITY IS MARKEDLY IMPROVED BY A PRETREAT-MENT WITH BENZALKONIUM CHLORIDE. SURFACES LIKE THIS ARE ONLY THROMBORESISTANT WHEN THE HEPARIN-BENZALKONIUM COMPLEX IS SORBED IN THE POROSITY.

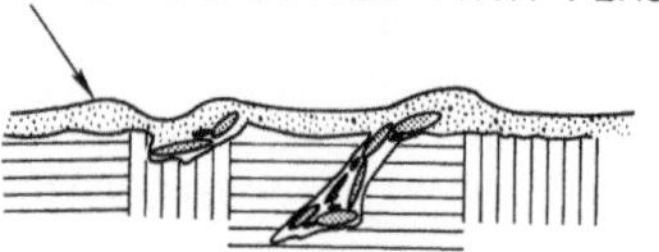

A LARGE FRACTION OF THE HEPARIN SORBED ON CARBON SURFACES IS ELUTRIATED RAPIDLY IN PLASMA, BUT A SMALL RESIDUAL AMOUNT PERSISTS.

Fig. 4.8 Results for rough and porous carbon surfaces. (Reprinted from *Chemistry and Physics of Carbon: A Series of Advances*, 9, 1973, with the permission of Taylor & Francis Group LLC-Books)

Where the Studies Led

The long and tedious study funded by the NIH proved to be a lesson in serendipity. The GBH coating was thromboresistant due to its porosity rather than to the presence of graphite. Any material with similar porosity might have worked the same way. The pure, smooth, clean LTI carbons were thromboresistant. Covalent bonding did not increase their thromboresistance; rather, it decreased the compatibility attributed to the presence of thrombogenic, unbonded, active sites.

Ultimately, it was Gott's experience with the use of carbon pigment paint that led to the discovery that isotropic pyrolytic carbon was thromboresistant.

In a review by Gott and his colleagues [22], the survival rate for the Gott-Daggett carbon-coated bileaflet valve (Ref. [23]) was updated to greater than 25 years. This was because the flexible silicone leaflets survived without deterioration.

If it hadn't been for the curiosity promoted by Gott's development of the GBH coating and the Gott-Daggett bileaflet valve replacement, General Atomic may never have discovered that its isotropic carbon had potential for use in heart valve replacements.

The General Atomic collaboration with NIH's Artificial Heart Program proved to be a productive learning experience. The melding of science and technology coming from the energy and medical fields produced a viable solution, not for the whole heart, but certainly for fabricating replacements for failing heart valves.

Fig. 4.9 All-carbon valve prototype developed at General Atomic. (Reprinted from the *Journal of Biomedical Materials Research* Part A. Correlations between blood compatibility and heparin adsorptivity for an impermeable isotropic pyrolytic carbon, MD Ramos, et al., 2004 with the permission of John Wiley & Sons)

During the course of the NIH study, General Atomic's President Frederic de Hoffmann requested that an *all-carbon tilting disc valve* be designed [24] and used in a demonstration of a new product that had bubbled up from de Hoffmann's John Jay Hopkins Laboratory for Pure and Applied Science (Fig. 4.9).

References

1. Gott VL, Koepke DE, Daggett RL, et al. The coating of intravascular plastic prostheses with colloidal graphite. Surgery. 1961;50(2):382–9.
2. Gott VL, Daggett RL, Koepke DE, et al. Replacement of canine pulmonary valve and pulmonary artery with a graphite coated valve prosthesis. J Thorac Cardiovasc Surg. 1962; 44(6):713–21.
3. Gott VL, Whiffen JD, Dutton RD, et al. The anticlot properties of graphite coatings on artificial heart valves. Abstract of paper submitted for presentation at the Sixth American Carbon Conference, Pittsburgh, PA, June 17–21, 1963.
4. Gott VL, Whiffen JD, Dutton RD. Heparin bonding on colloidal graphite surfaces. Science. 1963;142:1297–8.
5. Hastings FW, Harmison LT. Proceedings of artificial heart program, National Heart Institute, Hegyeli RJ, editor, June 9–13, 1969.
6. Bokros JC, Ellis WH. US Patent No 3,526,005, Carbon Medical Implant Devices, issued September 1, 1970.
7. De Laszlo. US Patent No 3,526,906, issued September 8, 1970.
8. Bokros JC. US Patent No 3,579,645, Broad Medical Implants, issued May 25, 1971.
9. Bokros JC. US Patent No 3,676,179, Residual Stress, issued July 11, 1972.
10. Bokros JC, Ellis WH. US Patent No 3,677,795, PyC Medical Devices, issued July 18, 1972.
11. Bokros JC, Ellis WH. US Patent No 3,685,059, Medical Devices, issued August 22, 1972.
12. Bokros JC, Ellis WH. US Patent No 3,707,006, Orthopedic Implant Devices, issued December 26, 1972.

13. Bokros JC. US Patent No 3,783,868, Percutaneous Device, issued January 8, 1974.
14. Bokros JC. US Patent No 3,971,134, Dental Implant, issued July 27, 1976.
15. Bokros JC. US Patent No 4,131,957, Finger Joint, issued January 2, 1979.
16. Bokros JC, Haubold AD, Akins RJ, et al. Trends in prosthetic heart valve design in. In: Bodnar E, Frater RWM, editors. Replacement cardiac valves. New York: Pergamon Press, Inc; 1991. p. 333–55.
17. Bokros JC, Haubold AD, Akins RJ, et al. The durability of mechanical heart valve replacements: past experiences and current trends in. In: Bodnar E, Frater RWM, editors. Replacement cardiac valves. New York: Pergamon Press, Inc.; 1991. p. 21–48.
18. Haubold AD, Bokros JC. Chapter 1: carbon in medical devices. In: Williams DF, editor. Biocompatibility of clinical implant materials, CRC series in biocompatibility, vol. 2. Boca Raton: CRC Press, Inc; 1981. p. 3.
19. Bokros JC, Gott VL, LaGrange LD, et al. Heparin sorptivity and blood compatibility of carbon surfaces. J Biomed Mater Res. 1970;4:145–87.
20. LaGrange LD, Gott VL, Bokros JC, et al. Compatibility of carbon and blood. In: Hegyeli RJ, editor. Proceedings of artificial heart program, June 9–13 1969. Washington DC: National Heart Institute; 1969.
21. Bokros JC, Gott VL, LaGrange LD, et al. Correlations between the blood compatibility and heparin sorptivity for an impermeable, isotropic pyrolytic carbon. J Biomed Mater Res. 1969;3:497–528.
22. Gott VL, Alejo DE, Cameron DEMD. Mechanical heart valves: 50 years of evolution. Ann Thorac Surg. 2003;76:S2230.
23. Gott VL, Daggett RL, Young WP. Development of a carbon coated central-hinging bileaflet valve. Ann Thorac Surg. 1989;48:528–30.
24. Bokros JC. US Patent No 3,546,711, Heart Valve, issued December 15, 1970.

Chapter 5
First Carbon Heart Valve Replacement

Michael DeBakey, a participant in the Artificial Heart Program, was impressed by not only the thromboresistance but also the durability of General Atomic's isotropic carbon. In addition to his work with the Artificial Heart Program, he was independently developing a new heart valve replacement for the aortic position.

The failure of silicone ball occluders had stymied work in this area, and DeBakey was determined to find a solution. Working through Surgitool, he placed a purchase order with General Atomic in 1965 for some hollow carbon balls.

Surgitool, a small Pittsburgh company owned by innovative mechanical engineer Harry Cromie, was then catering to a number of enterprising cardiothoracic surgeons. Among these were George Magovern (at the Allegheny General Hospital, Pittsburgh) and Arthur Beall (at the Baylor College of Medicine in Houston). All these researchers pursued heart valve replacement through the door opened by Gibbon's heart-lung machine, an innovation which made open heart surgery possible.

Cromie was a most gifted designer. For example, he designed and fabricated a ball-heart valve for George Magovern that did not use a cloth sewing cuff. Instead, it used a circular array of open curved metal needles held in place by a mechanism created by Cromie. When placed within an annulus, and actuated by the surgeon, the structure closed, grabbing the annulus wall and securing it firmly in place. This remarkable mechanism was made entirely of titanium.

Cromie wanted to simplify the method of fixation, lessening bypass time and reducing thrombus formation. This historic valve is worth mentioning because a patient who had the valve implanted in the 1960s was later admitted to a hospital for a failing mitral valve. His aortic valve was a Magovern-Cromie valve that had been implanted 42 years before and which was still functioning normally [1]. It was the longest functioning prosthetic valve of any kind ever documented. What a tremendous testament to the integrity of Harry Cromie's handiwork.

J. Bokros, *Heart of Carbon*, https://doi.org/10.1007/978-3-031-17933-4_5

Navigating Unforeseen Complications

When Cromie approached General Atomic, the business did not include carbon balls and certainly not hollow ones for use at the core of a heart valve. Because of our contract with the Artificial Heart Program, that Surgitool purchase order found its way to my desk. At the time I was working in the laboratory, developing carbon for coating fuel particles for use in the energy industry.

I was excited by DeBakey's interest. On the other hand, going down that path was at odds with General Atomic's corporate focus on the nuclear energy business. However, de Hoffmann's John Jay Hopkins Laboratory for Pure and Applied Science provided a suitably reasonable context for producing carbon components for heart valve replacements.

The issue of product liability then struck home with us. After a number of high-level meetings, de Hoffmann directed his lawyers to meet with Dr. DeBakey in Houston to discuss the terms and conditions associated with the use of its carbon product in human subjects.

A meeting was set. The attorneys and I traveled to meet with Dr. DeBakey at the Baylor Medical Center. On arrival, we were directed to DeBakey's private conference room. Before the meeting commenced, Dr. DeBakey appeared in his usual operating garb that included his famous white boots. Dr. DeBakey (Fig. 5.1), a commanding personality, addressed the lawyers:

> Gentlemen, my patients are very sick, sometimes dying before they are moved into the operating room. Even though your carbons may not be perfect, they are better than what I got.

I recall his verbiage clearly. On that note, the meeting that had yet to commence was cordially concluded.

As it turned out, the production of hollow carbon balls for use as an occluder (in a ball valve replacement for a diseased valve) was not straightforward.

First, although the deposition process (Fig. 5.2) was suitable for depositing thin carbon coatings on fuel particles, producing a coating 20 times thicker (such as was required for heart valve replacements) posed a challenge.

One of the first things to consider was which hydrocarbon would best serve as the carbon source to fabricate heart valve components. Although isotropic carbons

Fig. 5.1 Michael DeBakey MD. (Courtesy of Methodist Hospital Houston Texas)

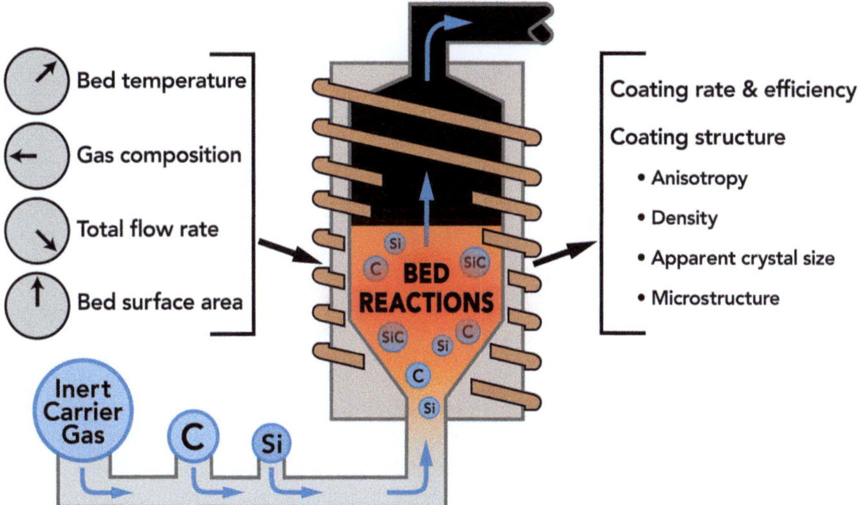

Fig. 5.2 High-temperature furnace used to deposit pyrolytic carbon. The four parameters on the left control the reactions in the fluidized bed that in turn determine the structure and properties of the carbon deposited. (Used with the permission of Artivion, Inc. and Kathleen Selbrede)

could be produced using methane, Ron Beatty at the Oak Ridge Laboratory in Tennessee had used propane to successfully deposit isotropic carbons at 1200 °C, with densities near 2.0 g/cm^3 at rates in excess of 200 μ/h. (Figure 19 on page 38 of Ref. [2]).

Richard J. Bard [3] also reported that dense isotropic carbon could be produced at 1200 °C using a number of different hydrocarbons, including propane, at very high rates. Accordingly, propane was selected as our carbon source, because it was readily available at high purity and low cost and coatings up to 1 mm thick could be deposited within a few hours.

In Fig. 5.2, the total surface area of the particles in the fluidized bed process is an important primary parameter that controls the structure of the carbon deposited. But the surface area of the bed is obviously continuously changing, as is the structure of the carbon being deposited.

Those who attempted to institute a steady state by establishing a material balance failed to achieve good control of the bed surface area [4]. Though the reasoning behind that strategy was correct, the process did not work very well because there was, at the time, no way to sense the total bed surface area in real time while the coating was in progress (Fig. 5.3a). The poor control of the bed surface area meant that the process was by no means a precise one, but rather a species of deficient processes termed "contraptions." A contraption is defined as a process positioned between an invention and a contrivance.

Before the year 1990, fluctuations in the bed area, like those depicted in Fig. 5.3b, caused significant fluctuations in the hardness and durability of the carbon

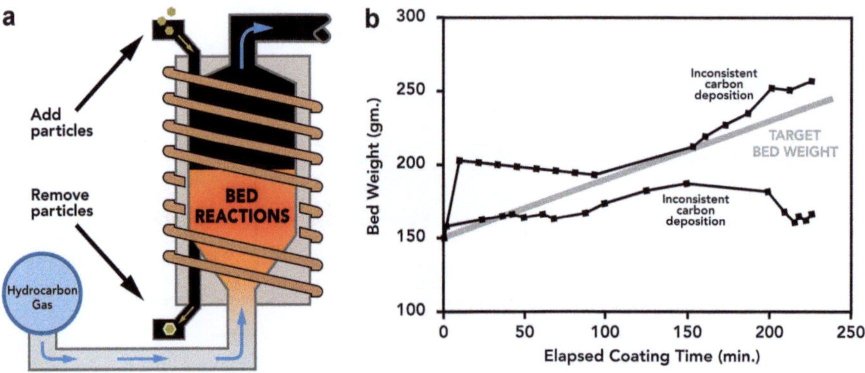

Fig. 5.3 (**a**) A process that attempted to reach a steady-state bed size with a material balance. It was presumed that if the surface area of the particles introduced from above and the surface area of the carbon introduced from below were balanced by the surface area of the coated particles extracted from below, a steady-state bed surface area could be achieved. (Reprinted from the *Journal of Heart Valve Disease*. Pure pyrolytic carbon: preparation and properties of a new material, On-X® carbon for mechanical heart valve prostheses, JL Ely, et al., 1998 7;6 with the permission of ICR Publishers Ltd). (**b**) Bed surface areas versus time for long coating times that were required to produce carbon-coated components for heart valve replacements. Results for two identical runs using guesstimated parameters sought the same target, but results were not reproducible. (Reprinted from the *Journal of Heart Valve Disease*. Pure pyrolytic carbon: preparation and properties of a new material, On-X® carbon for mechanical heart valve prostheses, JL Ely, et al., 1998 7;6 with the permission of ICR Publishers Ltd)

Hours	Thrombus formed in lumen		
	Pure silicon carbide	LTI carbon co-deposited with silicon carbide	LTI carbon without silicon carbide
2		◖◖◑○○	○○◑○○
336	●●◕	◖◑○○○	○◑○○○

LaGrange LD, Gott VL, Bokros JC, Ramos MD. Compatibility of carbon and blood. Hegyeli RJ, Editor, PROCEEDINGS, Artificial Heart Program Conference, Washington, DC, June 9-13, 1969, pp. 47-58.

Fig. 5.4 In vitro results for rings coated with impermeable pure silicon carbide, LTI carbon co-deposited with silicon carbide, and LTI carbon without silicon carbide. All were tested in canine venae cavae [5]. (Used with the permission of Artivion, Inc. and Kathleen Selbrede)

deposited using a propane precursor. The problem was corrected by co-depositing silicon as SiC, with the carbon solving the durability problem.

However, SiC had been shown to be thrombogenic (Fig. 5.4). Hence, only a small amount (~7 wt % Si present as SiC) was used. This amount of silicon was still thromboresistant, but the composite performance was not as good as it could have been by omitting it. Reducing the degree of anticoagulation therapy required post-surgery would not be possible.

Manufacturing the First Hollow Balls

The purchase order from Surgitool in 1967 was simple. It requested various-sized hollow balls with a specific density near to that of blood. The rest was up to us. The oil companies needed assurance that anything supplied had to be demonstrated safe and efficacious. The details of their manufacture are described at length in Ref. [5] (pp. 140–147).

The best method found to manufacture hollow carbon balls was to hollow out a solid graphite ball through a small hole drilled in its side. The core of the ball was removed using a special tool. To make the balls radiographically visible, a fine tungsten screen liner was rolled up and inserted through the hole. The hole was sealed by cementing a threaded plug in place. Baking the assembly converted the cement to graphite. The balls could then be coated with a single layer of Pyrolite Carbon, 0.5 mm thick.

The density of the ball was an important factor affecting its dynamic performance. Excessive inertial lethargy could retard the ball's forward and backward excursions, harming the resulting hemodynamics of DeBakey's valve design.

To determine the effect of ball density on flow, 12 balls with densities varying between 0.95 and 1.15 g/cm^3 were fabricated, to be used in vitro in test valves evaluated using a pulse duplicator developed by a graduate student at Baylor.

In advance of testing the DeBakey valve, we hypothesized that the likely optimum density would be near 1.05 gm/cm^3 (the density of blood). Our tests went forward in expectation that this hypothesis was correct.

The valves to be tested were shipped to the graduate student for a full hemodynamic evaluation.

The balls were hydrostatically tested to 15,000 psi (the limit of the test equipment). When loaded in compression between parallel steel plates, the fracture (first crack) occurred at loads ranging from 300 to 1000 pounds.

Wear testing produced a surprise. An accelerated tester using rates expressed in cycles per minute (cpm) was used:

1. At 100 cpm, pumping 6 l/min of distilled water and using 130/80 mm Hg pressure, the wear rates were measured using a microbalance by weighing the unit before and after testing, thereby delivering an accurate reading of 2 millionths of an inch, which translated to 0.00004 inches of wear per year.
2. At 500 cpm for 50×10^6 cycles, the wear rate was ten times faster.
3. At 2500 cpm, the wear rate was 3000 times faster than when measured at 100 cpm.

   The wear rates exhibited a strong functional dependence on the cycling rate.

In 1967, with the data in hand, the only option available was to use distilled water at 100 cpm and extrapolate that data to estimate the actual wear rate. The values were measured in millionths of an inch and extrapolated to 200 years, which was the time it would take to wear halfway through the 0.020-inch-thick coating (Figure 21 on page 145 of Ref. [5]).

The Valves that Took a Detour

Mystified after 6 months of waiting for results, I decided to ask the Baylor graduate student about his findings for the optimum density of the balls. His response was that he had never received the valves with carbon-coated balls we had shipped him. Dumbfounded, I suggested that he check with the shipping department to see if they were received, as our records indicated that their receipt by Baylor was duly acknowledged.

Checking with Dr. DeBakey's secretary, the graduate student reported back to me that Dr. DeBakey had surgically implanted every single one of those valves in patients.

Accordingly, we decided that despite having none of the expected density data in hand, we would move ahead anyway, trusting our intuition that the density of the ball *should* match that of blood. We hoped that the tests that were curtailed would have proven to be superfluous.

We needn't have despaired. The graduate student subsequently conducted the duplicator studies we desired. He determined empirically that valves with a hollow carbon ball that bore a neutral density (matching the density of blood) led to significantly less turbulence, reflected on both sides (upstream and downstream) of the valve (Fig. 5.5). The results were published in a discussion presented in the then-forthcoming *Proceedings* of the Second National Conference on Prosthetic Heart Valves held in 1968 in Los Angeles and edited by L. A. Brewer, MD [6].

The balls (Fig. 5.6a) were shipped to Surgitool in early 1968. The first DeBakey-Surgitool aortic valve was implanted in 1968 (Fig. 5.6b). From 1969 to 1980, 3300 ball valves were actually implanted. By that time, the curing problem was solved, and silicone balls, when properly cured, lasted indefinitely, making the use of carbon unnecessary. This of course neglects the advantage of having the same density as blood.

DeBakey-Surgitool Valves In Vivo

Between 1968 and 1978, there were approximately 20 reports of DeBakey titanium cage wear. Some fractures were described as insidious but asymptomatic, since the valves continued to function, despite being impaired, even if two of three struts had fractured [7, 8].

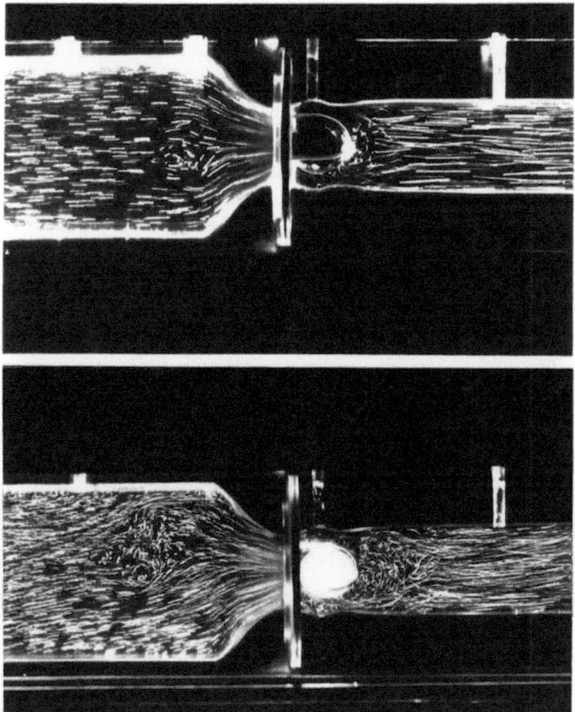

Fig. 5.5 Comparison of flow images for two ball valves, the top one with a black hollow carbon ball and the bottom one with a silicone ball as tested at Baylor. (Reprinted from Prosthetic Heart Valves ed. LA Brewer June 1968, courtesy of Charles C Thomas Publishers Ltd., Springfield, Illinois)

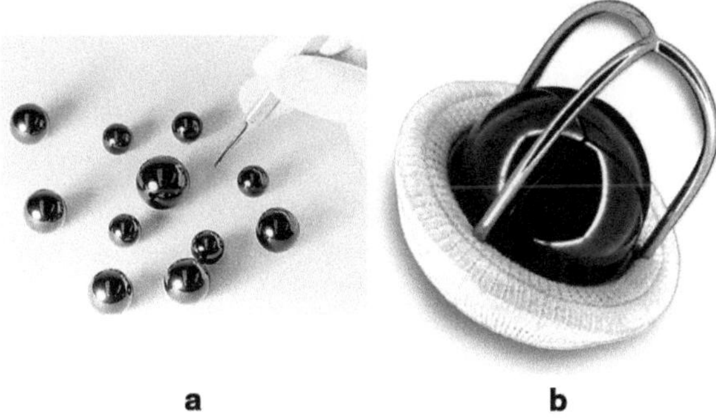

Fig. 5.6 (**a**) DeBakey hollow carbon balls. (Reprinted from the *Journal of Biomedical Materials Research* Part A. Correlations between blood compatibility and heparin adsorptivity for an impermeable isotropic pyrolytic carbon, MD Ramos, et al., 2004 with the permission of John Wiley & Sons). (**b**) The DeBakey-Surgitool heart valve with a hollow Pyrolite Carbon ball first used clinically in October 1968. (Reprinted from *Chemical and Mechanical Behavior of Inorganic Materials*, 1969, with the permission of John Wiley & Sons-Books)

Metallographic analysis indicated that the L-shaped bottoms of the three struts on a valve were susceptible to fatigue fracture traced to a preferred crystalline orientation induced in the rolling of the titanium sheets from which the cage was machined. It was found that if the sheet was cross-rolled at 90°, the texture could be reduced and the effect of the anisotropy mitigated.

In one case, a strut fractured, so the balls rubbed against it upon cyclical opening and closing. The titanium strut was ultimately worn to a sharp point by the Pyrolite ball.

In retrospect, if the base of the cage was formed with an inverted "T" rather than an "L," the stresses would have been symmetrically distributed, further reducing stress concentration.

Another patient, who already had a DeBakey aortic valve prosthesis that had been implanted in 1972, needed the mitral valve replaced. The DeBakey prosthesis was functioning normally after almost 34 years and is still providing good performance [9]. The mean aortic pressure was 20 mm, with no valve insufficiency or peri-prosthetic leak. The mitral valve was replaced. Intraoperative inspection indicated the replacement of the aortic valve was not necessary. Good function of both prostheses in the early post-operative period and at a 3-month follow-up was reported.

In another case, a DeBakey-Surgitool mechanical heart valve that was in place for 32 years was explanted for dysfunction related to tissue overgrowth, not for any failure of its components [10].

References

1. Zlotnick AY, Shiran A, Lewis BS, Aravol D. A perfectly functioning Magovern-Cromie sutureless prosthetic aortic valve 42 years after implantation. Circulation. 2008;117(1):e1–2.
2. Bokros JC. Chapter 1: Deposition, structure, and properties of pyrolytic carbon. In: Walker PL, editor. Chemistry and physics of carbon. New York: Marcel Dekker, Inc.; 1969. p. 5.
3. Bard RJ, Baxman HR, Berting JP, et al. Carbon. 1968;6:603.
4. Bokros JC, Akins RJ. US Patent No 3,977,896, Steady State Bed, issued August 31, 1976.
5. Bokros JC, LaGrange LD, Schoen FJ. Control of carbon structures. In: Walker Jr PL, Thrower PA, editors. Bioengineering, chemistry and physics of carbon, vol. 9. New York: Marcel Dekker Inc.; 1973. p. 103–71.
6. Proceedings of Second National Conference on Prosthetic Heart Valves. May 30-31, June 1, 1968 (Brewer III, LA, MD, ed) Charles Thomas, Springfield, IL 1969.
7. Scott SM, Sethi GK, Paulson DM, Takaro T. Insidious strut fractures in a DeBakey-Surgitool aortic valve prosthesis. Ann Thorac Surg. 1978;25(4):382–4.
8. Von de Emden J, Eberiein U, Breme J. Asymptomatic strut fracture in DeBakey-Surgitool aortic valves. Texas Heart J. 1990;17(3):223–7.
9. Beiras-Fernandez A, Oberhuffer M, Kur F, et al. 34-year durability of a DeBakey-Surgitool mechanical aortic valve prosthesis. Interact Cardiovasc Thorac Surg. 2006;5:637–9.
10. Butany J, Naseemuddin A, Feindel CM. DeBakey-Surgitool mechanical heart valve prosthesis, explanted at 32 years. Cardiovasc Pathol. 2004;13(6):345–6.

Chapter 6
First All-Carbon Mitral Valve Replacement

While we were designing and developing DeBakey's ball valve, Arthur Beall, Jr., MD of Baylor University in Houston, Texas, and a colleague of DeBakey, pioneered the search for a thromboresistant mitral valve replacement [1, 2]. The early Model 104 of the Beall mitral valve used a Dacron-covered sewing cuff, struts covered with Teflon tubing, and a Teflon occluder.

Both DeBakey and Beall were asking for a carbon-coated metal cage for mitral and aortic valves to eliminate the use of polymeric materials. It was not possible to coat titanium, stainless steel, or chrome-cobalt with Pyrolite Carbon, so we searched for a metal that could survive the high temperature (~1300 °C) required during carbon deposition, thereby making it possible to fabricate a carbon-coated valve cage.

Only one possibility was found. It was a refractory metal alloy of molybdenum (with rhenium added) that was proprietary, and which also had a single source willing to provide the alloy *only if* they were indemnified and held harmless, to which we agreed.

In 1969, I traveled to Houston, Texas, to meet with Arthur Beall and Michael DeBakey. DeBakey wanted to discuss a hollow carbon disc, pictured in Ref. [3] as Fig. 15a, for his envisioned mitral prosthesis and also an advanced aortic prosthesis composed entirely of carbon, with a carbon-coated metal cage, a hollow carbon ball occluder, and a carbon-coated orifice. I arrived with a prototype carbon-coated metal cage (Fig. 6.1).

Dr. Beall envisioned his all-carbon mitral prosthesis as a companion to DeBakey's carbon aortic prosthesis. DeBakey and Beall were both associates at Baylor but not in direct collaboration. I was in between the two pioneers. Harry Cromie, the entrepreneurial mechanical engineer mentioned in the previous chapter, was simply "going with the flow." He was soft-spoken, personable, and an astute businessman, a major factor in introducing carbon to the surgeons.

I arrived a day early so I could be in DeBakey's waiting room before 5:00 a.m., hoping to intercept him as he arrived. At just before 6 a.m., I was awakened by a burst of light as he stepped out of his office dressed for his first case, already

© The Author(s), under exclusive license to Springer Nature Switzerland AG 2023 45
J. Bokros, *Heart of Carbon*, https://doi.org/10.1007/978-3-031-17933-4_6

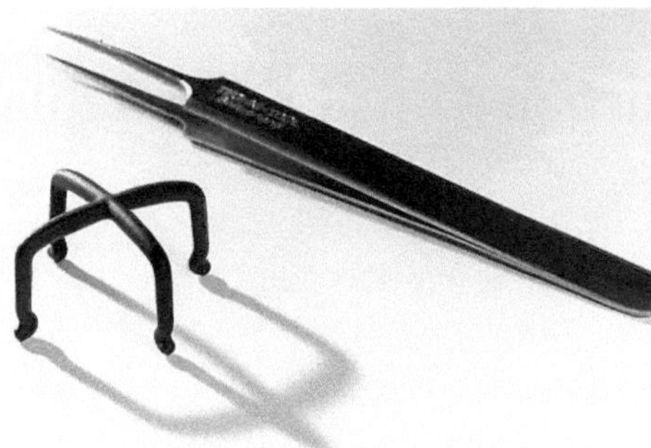

Fig. 6.1 Carbon-coated metal valve cage. (Reprinted from the *Journal of Biomedical Materials Research* Part A. Heparin sorptivity and blood compatibility of carbon surface, KD Vos, Ramos, et al., 2004 with the permission of John Wiley & Sons)

wearing his famous high-top white rubber boots. He knew I was coming but had anticipated meeting me between operations.

He told me that he normally started work at 4 a.m. to get his paperwork out of the way before he operated. He was terse concerning his interests and asked me to attend a later case to replace a mitral valve in an elderly lady (to start at 9 a.m.) and discuss his valves thereafter.

In the meantime, I met with Dr. Beall, who was also working with Harry Cromie's Surgitool.

Dr. Beall had envisioned an all-carbon mitral valve that had a carbon-coated metal frame and a solid carbon disc. I did not anticipate a competitive situation between Beall and DeBakey. My mission was to introduce General Atomic's carbon heart valve replacements, with all-carbon components, whether with DeBakey and/ or Beall. Cromie had made that possible.

At the time, General Atomic's President, Frederic de Hoffmann, proud of the firm's carbon technology for use in implanted devices, all protected by a "blanket" of patents, was concerned about product liability. Since Surgitool had become a subsidiary of Travenol, Inc., an established, financially secure medical device company, General Atomic's attorneys were comfortable. Travenol understood that General Atomic was the sole source for Pyrolite Carbon.

As you may have guessed, all of General Atomic's supply agreements required that not only was General Atomic not responsible for any sort of liability but it also required a solid "hold harmless" indemnification. General Atomic was to bear no responsibility and asserted that the purchaser as a device vendor was solely responsible. In any event, General Atomic was held harmless, as spelled out in rigid legalese.

This defensive attitude stemmed from oil companies' particular sensitivity to antitrust regulations with respect to price fixing. In fact, under no circumstance would General Atomic raise an existing price for carbon provided to a vendor. But if any vendor raised prices, General Atomic's corresponding price also rose, but less than the vendor's increase.

As it turned out, when 9 a.m. rolled around, I was busy with Dr. Beall when the building loudspeaker system suddenly announced, "Dr. Bokros, you are needed by

Dr. DeBakey." Arthur looked startled. "Hurry along," he urged me, "the boss is beckoning."

When I arrived at the operating theater, the head nurse stated with a glare, "When Dr. DeBakey says to be here at 9 a.m., you need to be here at 9 a.m.!"

The door opened. DeBakey's expression was hidden under his mask, but his eyes told me he was angry. Compounding the situation, he was experiencing difficulty with the operation.

Nervously I stood at his shoulder as he instructed me about the case. Afterward, I explained that I became squeamish at the sight of blood and was concerned about being in the way or stepping on important tubing and causing a problem. He told me not to worry, he understood, and he thus put me at ease.

He described the problem encountered with the patient and assured me with a grin that everything turned out okay. (To set the record straight, I have to confess that I was never squeamish at the sight of blood, but I felt thoroughly intimidated by DeBakey. It was the first excuse that came to mind. Fortunately, all was pardoned by his "okay.")

In the late 1960s (or possibly the early 1970s, I don't recall), I attended a conference about heart valves that honored pioneers in the field, including Albert Starr for his development of the Starr-Edwards ball valve replacement. I was sitting next to Dr. DeBakey, way in the back. Dr. Nina Braunwald was presenting her newly developed aortic ball type replacement, cloth covered valve cage, and a polymeric ball occluder.

At the end of her discussion, Dr. DeBakey rose to comment. As famous as he was, he was immediately recognized and was yielded the floor. He first complimented Braunwald for her efforts but then went straight to the point. "Nina, you have incorporated all of the bad features of ball valves into one valve."

I was startled to say the least by his devastating words. He made several further observations, concluding his critique with these words: "Looking forward, (showing Fig. 6.2) I'm showing you an advanced DeBakey-Surgitool aortic prosthesis."

Fig. 6.2 Prototype of an all-carbon version of the DeBakey-Surgitool aortic prosthetic valve. (Reprinted from *Chemistry and Physics of Carbon: A Series of Advances*, 9, 1973, with the permission of Taylor & Francis Group LLC-Books)

Fig. 6.3 Jack Bokros introduces Dr. DeBakey to the On-X valve. (Courtesy of Sherry Shaw–Smith)

With the exception of its sewing cuff, the new device had carbon coating on the metal cage, the orifice, and all the other surfaces.

Afterward, Harry Cromie pulled me aside and told me that DeBakey had chided him, saying "Harry, we should have cornered the carbon technology—gotten an exclusive right." I knew that to be completely unrealistic, as the jealously protected technology was Dr. de Hoffmann's pride and joy. Ironically, DeBakey's advanced model never made it to the marketplace. Dr. Nina Braunwald's valve suffered from the deficiencies mentioned by DeBakey and was withdrawn from the market.

The circle thus drawn would take several more decades before it was complete. The indefatigable Michael DeBakey would live long enough (99 years) to get a glimpse of the tremendous technological advance of the On-X valve made of On-X carbon (Fig. 6.3).

The Beall Valves Advance the State of the Art

In 1973, Beall reported on the background and development of his all-carbon mitral valve, manufactured by Surgitool Company, the same manufacturer for the DeBakey aortic valve (prior to Travenol's acquisition of Surgitool). It had the general features of the Beall-Surgitool mitral Model 104, with a Dacron-covered sewing cuff, struts covered with Teflon tubing, and a Teflon occluder. Unlike its predecessor, however, all the surfaces of Beall's new model, cage and disc, were coated with carbon [4, 5].

In the improved all-carbon Beall valve the struts and disc are surfaced with carbon. The construction details of the new Beall-Surgitool all-Pyrolite mitral valve, Model 105, are described in Ref. [6], pages 147–151.

Because of the uncertainty of using accelerated testing to determine the durability of the DeBakey aortic valve, the Beall Model 105 valves were tested in calves, extending out to 12 months of continuous monitoring for durability and function. The bovine results indicated that it would take at least 140 years to wear through the carbon coating (Ref. [2] and pages 147–151 of Ref. [6]).

The clinical testing of the Beall Model 105 valve was initiated in June of 1971. Two reports in 1973 indicated that the Model 105 provided extremely low incidence of thromboembolic complications, even without anticoagulation [2, 4]. The reports included 175 cases monitored over a 16-month period. There were no deaths related to the prosthesis. Four thromboembolic episodes were reported; the only fatal one was an embolus from the left atrial appendage that had obstructed the prosthesis.

The prosthesis was announced as available for general use in 1973. This was before federal legislation of medical devices was enacted, which would go into effect in 1976. An approval by the Federal Food and Drug Administration (FDA) was not required.

A Major Wake-Up Call

An addendum to the report was submitted after the manuscript of Ref. [4] had been received and was at the press. That addendum disclosed the details of two deaths caused by the fracture of struts, thereby releasing the discs into the ventricular cavity.

It was concluded that the coating had been cracked during handling of the prosthesis. This assessment prompted a warning that "gentle handling" is mandatory: provisions must be made to insure that bending stresses are never applied to the struts during implantation. If the prosthesis is dropped, or otherwise mishandled, the valve must be returned to the manufacturer.

This was a wake-up call regarding the handling of prosthetic valves. Up to this point, no special attention was paid to packaging and handling. Now, valves needed to be packaged in a rigid container that immobilized the device, and they required a handle (whether separate or attached) so that sutures could be placed without directly grasping the valve.

Further, the user instructions stated that the valve must not be released from its holder until after the valve had been securely sutured in place. If the device required rotation, then the supplier needed to provide an appropriate device that could engage the valve after suturing, without stressing the valve cage. At this point in time (circa 1975), federal regulations of medical devices were already in draft form and would soon be vigorously enforced.

There were six instances reported of fatigue fractures of the cage, with associated escape of the disc. As a precaution, the thickness of the carbon coating on the metallic cage was increased, and packaging for the device provided, so the valve could be sterilized *in the package*, protecting the prosthesis until implantation. The improved valve was labeled Model 106 [5].

There were no further episodes reported of strut fracture. Overall, there were 18,000 Beall Models 103 and 104 valves distributed. Up until 1977, Travenol had sold 3000 Models 105 and 106 Beall valves until the Model 106 valves were sold to the Medical Engineering Corp of Racine, Wisconsin. An additional 1174 Beall Model 106 valves were sold up until August of 1982 [7].

One Patient's Three-Decade Track Record

A report exists of a Beall Model 106 mitral valve that had remained implanted for 30 years [8]. The woman received the valve at age 21. Thirty years later, the explanted valve showed massive pannus formation (pannus is scar tissue), extending over the sewing cuff on both the inlet and outlet sides, thus preventing complete valve closure. The patient presented a genuine valve card, dated 7 September 1976, stating that the prosthesis was a medium-sized Beall-Surgitool Model 106 valve.

The valve was replaced with a 31 mm On-X Life Technologies mitral prosthesis. The patient recovered uneventfully. At her last clinical visit (January 2010), the patient was asymptomatic. The On-X valve was functioning normally.

We planned to continue following this woman, sustained for 30 years with an all-Pyrolite Carbon valve and now rescued by a more advanced On-X carbon valve, a valve with improved carbon and superior hemodynamic function. One of the implanters was a good friend, Dr. Szabolcs Szentpetery of Hungarian heritage, who participated in the On-X clinical trials.

At the Dawn of the Industry

The state of affairs with respect to the development of prosthetic heart valves at the time can be found in two separate conference proceedings from that era. The first was compiled from the Conference on Prosthetic Valves for Cardiac Surgery held in Chicago, Illinois, September 9–10, 1960 (proceedings edited by Merendino AK, MD, and published in 1961) [9]. The other was the Proceedings of the Second National Conference on Prosthetic Heart Valves held in Los Angeles, California, May 30, May 31, and June 1, 1968 (conference chaired and proceedings edited by Brewer III LA, MD) [10].

The first proceedings had no reference whatsoever to carbon as a component in prosthetic cardiac valves. The second proceedings had but one reference in a discussion of a valve testing paper in Chap. 16 on page 254 of those proceedings [10] by a Baylor graduate student. The contribution described the comparative hemodynamic performance of two valves, the Starr-Edwards versus the DeBakey aortic valve. The DeBakey valve flow exhibited superior performance in this test. Incidentally, in those same proceedings, evidence for the failure of a silastic ball was depicted (in Fig. 222 in Ref. [10]).

In 1968, General Atomic was recognized with an award for the manufacture of the hollow carbon balls for the DeBakey valve, deemed to be one of the most important new products of the year (1971) [11]. Pyrolite Carbon was registered in 1966 [12].

Visitors were often escorted through the General Atomic laboratories to observe a small pump circulating an aqueous solution through an apparatus that included an all-carbon tilting disc valve that was designed and patented in 1968 (Fig. 4.9). It was not only the first all-Pyrolite Carbon valve, but it was also the first such valve where the occluder could open completely to a full 90° (parallel to flow). This innovative valve was animal tested but never used clinically.

By the mid-1970s, 100% of my focus was on the Medical Products Division, where I served as its Director. As such, my focus on nuclear energy ended. Summaries of my efforts developing nuclear fuel are in Refs. [13–17].

The carbon-coated uranium-thorium carbide particle fuel developed at General Atomic described in Ref. [17] was first used in the Fort St. Vrain Power Station located in Fort St. Vrain, Colorado. On December 11, 1976, the Fort St. Vrain power station first started to feed the grid with electricity generated using nuclear power.

Commercial operation for the Fort St. Vrain Power Station began on August 1, 1979. The reactor was refueled in July, 1981, and operated at 100% power. The Fort St. Vrain Power Station reactor set a new record for total net generation during a single month (160,184 megawatts for March, 1988). Due to continuing operational problems, however, the Public Service Company made the decision in December 1988 to shut down the Fort St. Vrain Power Station by June 30, 1990.

Although the facility was decommissioned, the HTGR that was in service at the Fort St. Vrain Power Station was the first of its kind, with potential well beyond the capabilities of conventional pressurized steam or boiling water reactors of the kind developed by General Electric and Westinghouse.

References

1. Beall AC Jr, Bloodwell RD, Liotta D, et al. Elimination of the sewing ring-metal seat interface in mitral valve prostheses. Circulation. 1968;37(Suppl II):184.
2. Beall AC Jr, Bloodwell RD, Arbegast NR, et al. Chapter 21: Mitral valve replacement with Dacron-covered disc prosthesis to prevent thromboembolism: clinical experience in 202 cases. In: Brewer III LA, editor. Prosthetic heart valves. Springfield, Ill: Thomas; 1970.
3. Bokros JC, Gott VL, LaGrange LD, et al. Heparin sorptivity and blood compatibility of carbon surfaces. J Biomed Mater Res. 1970;4:145–87.
4. Beall AC Jr, Morris GC Jr, Noon GP, et al. An improved mitral valve prosthesis. Ann Thorac Surg. 1973;15(1):25–34.
5. Beall AC Jr, Morris GC Jr, Howell JF Jr, et al. Clinical experience with an improved mitral valve prosthesis. Ann Thorac Surg. 1973;15(6):601–6.
6. Bokros JC, LaGrange LD, Schoen FJ. Control of carbon structures in Bioengineering. In: Walker Jr PL, Thrower PA, editors. Chemistry and physics of carbon, vol. 9. New York: Marcel Dekker Inc.; 1973. p. 103–71.

7. Fernandez J, Chang K, Gooch A, et al. Anatomic and clinical analysis of 96 Beall prostheses explanted over a 13 year period. Chest. 1983;83:632–7.

8. Topaz O, Rutherford MS, Mackey-Bojack S. Beware of the B(e)all valve. Tex Heart Inst J. 2010;37(2):237–9.

9. Proceedings of Conference on Prosthetic Heart Valves for Cardiac Surgery Sept 9–10, 1960 (Merendino AK, ed) Charles Thomas, Springfield, IL 1961.

10. Proceedings of Second National Conference on Prosthetic Heart Valves. May 30–31, June 1, 1968 (Brewer III, LA, MD, ed) Charles Thomas, Springfield, IL 1969

11. IR-100 Award. Special carbon for artificial heart valves. Industrial Inc.; 1971.

12. Pyrolite, serial no 312,673, filed Nov 20, 1968, Carbon coated prosthetic devices, heart valves and pumps, vascular tubes and hip joint balls, class 44 (int. Cl. 10) First use August 6, 1968, in commerce August 6, 1968.

13. Bokros JC, Guthrie GL, Schwartz AS. The influence of crystallite size on the dimensional changes induced in carbonaceous materials by high temperature radiation. Carbon. 1968;6:55.

14. Bokros JC, Guthrie GL, Dunlap RW, Schwartz AS. Radiation-induced dimensional changes and creep in carbonaceous materials. J Nucl Mater. 1969;31:25–47.

15. Bokros JC, Dunlap RW, Schwartz AS. Effect of high neutron exposure on the dimensions of pyrolytic carbons. Carbon. 1969;7:143.

16. Bokros JC, Koyama K. Interpretation of dimensional changes caused in pyrolytic carbon by high-fluence neutron irradiation. J Appl Phys. 1970;41(5):2146.

17. Goeddel WV, Bokros JC. AIME nuclear metallurgy symposium. Delavan, Wisconsin, October 1966. In: Holden AN, editor. High temperature nuclear fuels, vol. 42. New York: Gordon and Breach; 1966. p. 85–104.

Chapter 7
Monoleaflet Tilting Disc Valves

Three valve companies (Medical Incorporated, Shiley Laboratories, and Kastec Inc.) were to begin producing six different versions of tilting disc valves in the 1970s. Because this evolution will be laid out chronologically, the reader will note that some overlap between different corporate trajectories simply can't be avoided.

My first exposure to the Lillehei-Kaster tilting disc valve came from a photograph published in an early 1970s trade magazine which depicted "a new valve for the failing heart." Robert (Bob) L. Kaster was an electrical engineer working in Dr. "Walt" Lillehei's laboratory at Cornell University.

My first conversation with Kaster focused on the polymeric disc. Kaster was not aware of General Atomic's isotropic pyrolytic carbons, but he was certainly aware of the deficiencies of polymers (e.g., Delrin) that had been used in the original Björk-Shiley valve before they were later replaced by Pyrolite Carbon in 1971.

I explained to Kaster that isotropic carbon was impermeable, was durable, and had already been proven for two valves developed for Drs. DeBakey and Beall. Kaster was pleased to hear this. He asked if discs coated with isotropic carbon could be supplied for him to test, and he invited me to visit his Cornell laboratory to discuss their new valve.

Soon thereafter, we delivered the discs that Kaster had requested, manufactured to his specifications.

An Unforeseen New Problem Arises

Several months later, I visited Cornell and was ushered directly into Kaster's lab. An actual test was in progress. Kaster's accelerated testing apparatus (one he had designed and built himself) was running at high speed.

Kaster dimmed the lights. Using a strobe light, he revealed clouds of bubbles emanating from the valve. "Look at that," he said. "What do you think?"

© The Author(s), under exclusive license to Springer Nature Switzerland AG 2023
J. Bokros, *Heart of Carbon*, https://doi.org/10.1007/978-3-031-17933-4_7

I had never seen evidence of cavitation in prosthetic valves before and was quite amazed and, frankly, concerned.

Kaster returned the disc under test to me. I in turn gave the disc to Frederick Schoen (PhD, Cornell University, Materials Science). I first met Fred during a lecture presented by me at Cornell before his graduation. After graduation, he was hired to become a staff associate in the Medical Products Division of General Atomic. Fred went ahead and examined the disc using scanning electron microscopy (SEM).

The SEM in Fig. 7.1 is astonishing. The central portion of the disc is "laced" with micro "canyons." The inner regions are severely eroded, while the outer surface is free of detectable damage. The damage was identified by Fred as cavitation damage that occurred at 2500 cycles per minute in the Kaster tester. This was the first solid evidence of cavitation in a heart valve replacement.

This discovery proved to be an early warning to designers of carbon heart valve replacements. At the time, cavitation was considered to be an artifact of accelerated testing. The reality was that the threshold for cavitation was closer to physiological conditions in vivo than previously thought. In the mid-1980s, however, cavitation had yet to be actually detected within a patient.

The surface of a functioning carbon heart valve replacement can be eroded by cavitation that occurs momentarily (and repeatedly) at points where hemodynamic forces reduce the vapor pressure below the ambient vapor pressure. The implosion pressure of the collapsing cavitation bubbles, calculated to be enormous, can produce catastrophic impacts that form micro-fractures visible as "canyons" in the carbon surface (Fig. 7.1).

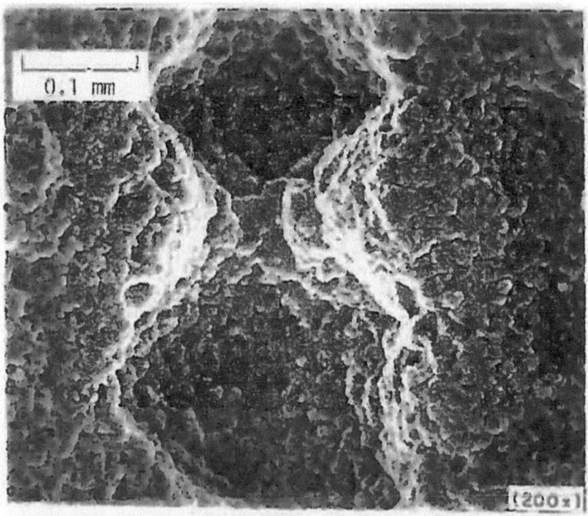

Fig. 7.1 Cavitation erosion in an accelerated test (2500 cycles per minute). Cavitation was detectable by the clouds of bubbles emitted on opening and closing of the valve observable with the aid of a strobe light. (Reprinted from *Chemistry and Physics of Carbon: A Series of Advances,* 9, 1973, with the permission of Taylor & Francis Group LLC-Books)

As you might imagine, the devastation of cavitation can wreak havoc on blood cells. Hemolysis due to cavitation has been reported as fulminating.

Cavitation, like the plague, must be avoided.

Business Complexities Take Center Stage

Kaster actually conceived his valve concept in 1966 and 1967 and applied for a patent while employed at the University of Minnesota. In 1967, Lillehei accepted the prestigious chair of Professor of Surgery at Cornell University Center and Surgeon-in-Chief at New York Hospital. The Kaster and Lillehei US Patent 3,476,143, issued in 1967, was destined to become a core patent located at ground zero of a series of subsequent intellectual property disputes.

In 1967, the University of Minnesota granted a license to manufacture heart valves to Washington Scientific Industries. With Kaster's assistance, Washington Scientific Industries developed the Lillehei-Kaster heart valve, a clinically and commercially acceptable version of the Kaster patent.

In 1971, Washington Scientific Industries sold its heart valve business to Marshall Kriesel, who formed Medical Inc. shortly thereafter. The sale required Medical Inc. to become a party to, and pay all costs of, a lawsuit that had been brought in 1970 by Washington Scientific Industries and the University of Minnesota against Shiley Laboratories for alleged infringement of the Kaster patent by the Björk-Shiley valve. Costs were to be offset against royalty payments.

The suit was later litigated. The court found that the Björk-Shiley valve patent did not infringe the Lillehei-Kaster patent. The case was finally settled in 1975.

Shiley Laboratories moved forward with the Björk-Shiley valve using a Delrin disc, until they converted to Pyrolite discs in 1971. Kaster had adopted the use of carbon discs earlier, prior to the valve becoming a commercial product to be manufactured by Washington Scientific Industries.

Large Black Smudge on Laboratory Exterior Wall

In the years 1970–1973, General Atomic's Medical Products Division was located in a large circular laboratory (Fig. 2.3) configured as a pilot production facility for producing medical carbons in support of our contract with NIH's Artificial Heart Program (Chap. 4), as well as producing components for the DeBakey and Beall valves (Chaps. 5 and 6). The pilot facility also produced discs for Medical Inc.'s Lillehei-Kaster valve.

Progress was interrupted in 1970 when Dr. de Hoffmann drove into the back entrance of General Atomic and noticed a large black smudge on the side of the main circular laboratory building. On checking, he was informed, "That's Bokros's facility for producing carbon coated heart valve components for the Medical

Products Division." The smudge was carbon soot. The Medical Products Division was de Hoffmann's pet project.

Dr. de Hoffman invited me into his office, speaking politely in his distinctive Austrian accent.

"Dr. Bokros, I'm very sorry for the inconvenience, but you must move your pilot facility out of the laboratory. You're expanding. We have a nice building for you in Sorrento Valley. I'm very proud of the Medical Products Division. I'm sure you will be very pleased." He didn't breathe a word about the smudge to me.

This was excellent news because the Sorrento Valley facility was much more spacious and was equipped with all the power we needed. We thought we could complete the move in a couple of months. It actually took us 4 months to relocate the Medical Products Division.

Shortly after the move, when we were up and running smoothly, there came a "knock on the door." It was an Italian biomedical engineer. At the time, the Italian government was supporting projects spun off from its nuclear activity associated with the Euratom program (a joint European effort to develop nuclear power).

The Italian biomedical engineer told me that the heart valve business "looked interesting" and so he requested a tour of our facility.

You can well imagine my response! I explained to the Italian biomedical engineer that the Medical Products Division was a private entity, privately supported, and I politely escorted him to the door. The Italian biomedical engineer continued his work, ultimately producing an Italian "Björk-Shiley knock-off."

Our patents on isotropic carbon were not recognized in Italy because, in order to be valid, the patented item had to be produced there. By the same token, the Italian biomedical company was unable to distribute the valves using their own version of Pyrolite Carbon components outside of Italy until the Pyrolite Carbon patents expired (excepting those nations where General Atomic's patents had not been issued, such as China, Russia, and India).

An Invited Visit to the Italian Company Facility

In 1972, I responded to an invitation from the Italian biomedical company engineer to visit their facility in Italy. This represented an opportunity to sell Pyrolite discs to the Italian biomedical company for their "Björk-Shiley" valve.

We met in a large conference room with the head of the unit and his team of engineers. After a conversation about Italian biomedical company's valves, I offered to sell discs to them for about $55 US. The room became very quiet: a complete non-response to what I had just put on the table.

After the meeting adjourned, the head of the unit pulled me aside and asked if I had noticed how the engineers' jaws had dropped at our $55 price point. I answered no, I hadn't. He replied that he *had* noticed, and he pointed out why they were shocked: the Italian biomedical company's cost to fabricate a disc was about $125 US.

The unit head further elaborated that no matter where we set our price, they would never buy our discs, because the Italian biomedical company was a government project. "You're beating a dead horse," he explained.

Shortly thereafter, another business interest emerged. An executive of a medical company arrived at the General Atomic facility (presumably to do some "fishing").

I did not offer the executive a tour, only a seat at my metal desk in General Atomic's engineering building. I let him drive the conversation.

The executive pulled up his chair to put his feet up on my desk. He made it clear to me that the company would not buy "your carbon balls. I can make silicone occluders for about five dollars, and if I needed balls or anything made of carbon, I would just buy *you*."

His arrogant proclamation having gained no traction with me, we exchanged some pleasantries before he departed.

Marshall Kriesel and Medical Incorporated

Our immediate business was with Medical Incorporated's President, Marshall Kriesel, a large man often considered intimidating, a strong-willed and aggressive individual. Kriesel, with his attorney, if memory serves, formed a dynamic duo. They had acquired a large, beautifully suited home for Medical Inc. in 1971. Their facility, which had clean rooms and lacked for nothing, was located in a rural farming area near Minneapolis (which is why it was dubbed Medical Inc.'s "laboratory in a cornfield").

The team managing General Atomic's Medical Products Division made multiple trips to Marshall Kriesel's elaborate facility to discuss ongoing technical issues.

Marshall and his attorney were superb hosts, providing dinner at two of Minneapolis's finest restaurants—Manny's and Charlie's. I often wondered if the extravagance of those dinners was motivated in part by an underlying intention to acquire General Atomic's Medical Products Division in the future.

I could not help but be impressed by Charlie's Steak House. Not only were the steaks superb, but the urinals in the men's room were cooled by buckets of ice cubes kept full and fresh by a tuxedo-wearing attendant! That was a new one for me. We never experienced anything that ostentatious with the CEO and President of Shiley. (His prudence may have been governed by an accountant's instinctive frugality.)

Over time, Shiley's materials problems continued to plague their valves. Most of these problems originated in the Delrin disc.

In the meantime, Bob Kaster oversaw the Lillehei-Kaster project at Medical Incorporated. Kaster was involved for a time but, following his instinct for independence, formed Kastec Inc. in 1975, funded by a Norwegian entrepreneur who recognized the heart valve replacement market as a budding field and was on the lookout for opportunities. Earlier at Cornell (perhaps in 1974), Kaster phoned and asked if General Atomic could provide a disc with a hole located *precisely* in the center (emphasis on *precisely*). He was quite secretive concerning the purpose for the central hole. His inquiry was for a new heart valve as described in the next paragraph.

The valve was unique among the array of tilting disc valve replacements, a credit to Kaster's ingenuity. It had a long curved strut that extended through a hole in its center that guided the motion of the disc during operation. I considered it to be ingenious, albeit a bit noisy. The strut was titanium but was thought to carry a fairly small load during valve function, thus avoiding the fatigue fractures observed earlier in the Björk-Shiley valve.

On October 24, 1979, Kaster filed a patent application for a pivoting disc heart valve that would make its first appearance as the Hall-Kaster valve. Ultimately, it evolved into the Medtronic-Hall valve, after Medtronic, Inc. acquired Kastec in 1979, with General Atomic as supplier of the disc to Medtronic, Inc.

In the meantime, Medical Inc., having developed a close working relationship with General Atomic, expressed concern about relying on General Atomic's Medical Products Division as a sole source for Pyrolite Carbon discs. Medical Inc. asked General Atomic for a license to produce its own carbon for its valve products.

This request served notice to us that Kriesel's conspicuous hospitality was motivated by the desire to acquire a license to use Medical Products Division's carbon technology. As was true for all previous inquiries by others, General Atomic denied Medical Incorporated's request for a license. Kriesel's ambitions in this regard will be revisited later in this chapter.

Shiley Laboratories Develops Its Valve Designs

In the early 1960s, Don Shiley, a chief engineer of Edwards Laboratories, had proposed using a disc rather than a much larger and heavier ball to regulate flow in their lab's prosthetic valve. When Edwards failed to encourage development along those lines, he parted company and founded Shiley Laboratories, Inc.

In 1968, Shiley enlisted Dr. Viking O. Björk, professor of surgery at Stockholm Hospital, as a collaborator. To head the management slate, he hired an accountant to serve as Chief Executive Officer and President. An engineer was hired as Vice President of Operations.

We knew that Shiley had originally ignored carbon for valve applications, preferring Delrin, a polymeric material. He hadn't recognized that Delrin was permeable, was dimensionally unstable, and was not as durable as Pyrolite Carbon. Delrin was, however, less expensive than Pyrolite Carbon. At the time, Shiley had selected Delrin as an occluder. Pyrolite discs were the only available alternative to Delrin, but were priced higher.

General Atomic refused Shiley's request for a license to permit his laboratory to produce its own carbon discs.

Shiley's Vice President of Operations admitted that he had attempted to produce pyrolytic carbon in their facility in the early 1970s. Laughing at the recollection, the Vice President of Operations recalled that for all the trouble they took, they "only managed to fill the building with black smoke."

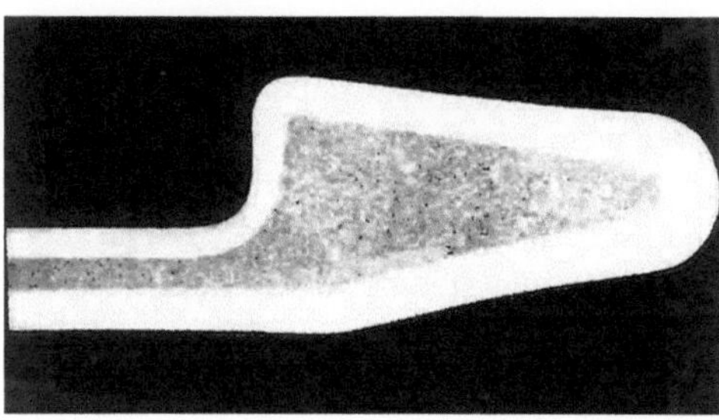

Fig. 7.2 A section through one of the first Pyrolite Carbon-coated graphite discs fabricated for the Björk-Shiley valve. (Reprinted from *Chemistry and Physics of Carbon: A Series of Advances*, 9, 1973, with the permission of Taylor & Francis Group LLC-Books)

In 1971, the original Björk-Shiley valve was modified, changing the disc from Delrin to Pyrolite Carbon while retaining the original plano-concave shape (Fig. 7.2).

Unlike Medical Inc., Shiley Laboratories and General Atomic were never close. Shiley's President and CEO remained perpetually aloof from us. The only interaction occurred when he visited me at General Atomic to ask for a discount. His price of a Delrin disc was less than a few dollars, whereas carbon discs were $55.

I listened to a litany of reasons why I should yield and reduce the price. Sensing a confrontation, I decided to hold back and listen. I knew better than to try to win an unwinnable debate with the CEO. Instead, I calmly told him, "The price is $55, and I cannot change that." There was a long pause as he gathered his papers and left without a word.

In 1975, Shiley Laboratories published a letter in *Circulation* 76;53:206 [1] that disclosed three strut fractures isolated to three production lots, triggering the issuance of a recall of all remaining valves from those lots that hadn't yet been implanted. Superficial analysis alleged those fractures to be isolated and restricted to those lots. The clinical use of the valve continued unabated.

In that same time frame, Shiley and General Atomic were obtaining scanning electron microscopy service from a laboratory in Orange County, near the Shiley facility. Dr. Fred Schoen and I were in the lab with two Shiley engineers. We watched the two stare at a Björk valve housing that displayed details of the structure. I took notice and reminded Fred that our observations had to remain confidential.

Medical Inc. Designs an Improved Valve

Medical Inc. and Shiley were fierce competitors because of the alleged patent infringement of the Björk-Shiley valve that had been filed before Medical Inc. existed. That competition escalated as Kriesel moved forward with Medical Inc.'s new, improved Lillehei-Kaster design.

In the early 1970s, Kriesel invited the Medical Products Division management team to a meeting in Minneapolis. The discussion initially focused on the shape of the disc. The new valve was to have a curved shape, rather than a flat disc used in the existing Lillehei-Kaster valve. Its curvature was spherical. It would have the same isotropic graphite substrate that had been used in all preceding valves, except for the addition of a small amount of tungsten (to improve its x-ray opacity to facilitate future diagnosis via remote imaging).

Late in the afternoon, Kriesel called a break in the meeting to announce a competition to provide a name for his company's new valve. The prize for best name was to be a case of wine. I submitted the prefix "Omni." It proved to be the winning name.

Kriesel took us to dinner at Charlie's. I wondered if steaks had been substituted for the promised case of wine, but the possibility of an involuntary swap was not discussed.

The Omniscience valve is pictured in the book *Heart Valve Replacement and Reconstruction* by Peter Starek [2], where the valve is compared to others. Photographs of the early Lillehei-Kaster valve (Fig. 18.5), the Björk-Shiley valve (Fig. 18.4), the Omniscience valve (Fig. 18.6), and the Medtronic-Hall valve (Fig. 18.7) are found on pages 224 and 225 of Starek's book. All are stenotic, resisting blood flow when compared with native valves. In fact, at the time of this writing, all of the above mechanical valves are less stenotic than bioprostheses. None, however, matched the hemodynamics of native valves.

The Omniscience valve was well accepted. Early and later results were judged to be satisfactory or excellent [3–5].

It took several years, and many visits, after the 1978 introduction of the Omniscience valve, to convince Kriesel that it was possible to redesign the valve's titanium orifice for manufacture with a Pyrolite Carbon coating. DeBakey and Beall were the first to achieve all-carbon status. We were certain that all-carbon status would benefit Medical Inc.'s Omni series of valves.

The challenge was to design the carbon orifice so that it was flexible enough to allow the disc to be inserted, yet stiff enough that the disc could never escape containment.

The problem was solved by using a cleverly designed fixture. The fixture could deform the orifice elastically enough for disc insertion. Once the disc was in place and the fixture removed, the disc could not escape. The details retained by Medical Inc. remain a trade secret.

Pyrolytic carbon is unique among metal and ceramic materials. It possesses a high strength and a low elastic modulus so that the fracture strain is unusually high (ranging up to 2% while fully elastic in that range without suffering permanent deformation). Strength and elasticity usually go hand in hand: as the strength increases, so does the modulus of elasticity.

General Atomic had originated the stretching assembly capability, promoting it in a widely distributed *Technical Bulletin* detailing the approach. Consequently, that technique could not be patented and remained in the public domain.

In 1983, Kriesel visited our facility in Sorrento Valley. When we were alone, he pulled up his chair, gathered up his massive hulk, and sat down. Pausing to open a book of matches, he lit one and slowly brought it up to his lips.

Looking me in the eye, he blew out the burning match. He paused again for effect and said simply, "Life is fragile."

It was an unforgettable moment, to be sure. I realized he had something on his mind.

He then revealed his intention to "hit the market" with the Omnicarbon valve. He needed 800 valve sets, to be delivered as soon as possible. Bob Akins clapped and left to do just that—800 Omnicarbon valves—wow! A production plan for 800 valve sets was put in place, with staged deliveries to be made ASAP.

It turned out that Kriesel had simultaneously undertaken a secret project to produce Medical Inc.'s own homegrown all-carbon valve. His 800 valve order with us was simply to stockpile enough inventory to last until Medical Inc. could produce its own valve. He ultimately accomplished this in 2003.

I neglected to confirm our normal 30-day terms and came to realize the purpose of Kriesel's "match act" when no payments had been received several months into the project! When payments were demanded, Marshall Kriesel's sad stories fell on deaf ears at General Atomic. Terms reverted to cash in advance until the funds past due were paid.

Recall that I had submitted the winning name "Omni" to win Kriesel's contest. The prize was to be a case of wine; "The best," he had promised. The prize never came!

I was both amused and upset by stories emanating from Marshall and his attorney about acquiring a load of valves from us with no money down. It felt too much like the time the Marshall/attorney team bragged about buying their laboratory in the cornfield "for a song."

I suppose this was all in good fun—if you're on the exploiting end of the deal. I have to admit that, despite their antics, I enjoyed their company. The 800 Omnicarbon valve sets were delivered, and future orders were to be cash in advance. Our friendly relationship survived.

There was a lull before "cash in advance" deliveries began. The demands for Omnicarbon components increased, and Medical Inc. was allowed to go back to normal payment terms. At the same time, I was aware that Medical Inc.'s secret program to produce its own carbon was shifting into high gear.

In the late 1970s, when the Medical Products Division of General Atomic was for sale, an agent representing a "foundation for responsible parenting" offered $5.5 million. We later discovered that *Kriesel had sent the agent.*

Medical Incorporated's Omnicarbon Valve

The Omnicarbon valve was introduced outside the United States in 1984. After this point in time, our personal relations went silent. Kriesel left and several new employees were hired from St. Jude Medical. Anecdotal hearsay regarding the clinical performance of the Omnicarbon valve outside the United States was excellent.

In July 2001, the Food and Drug Administration (FDA) granted approval for the valve to be used in the United States, without additional US patient trials. This approval was based only on foreign data, "unprecedented in the history of the FDA." More than 30,000 Omnicarbon valves had been implanted in other countries before approval in the United States.

In November 2001, Medical Inc. went public, trading under the symbol MDCVU and reportedly raising $6.7 million dollars. Kriesel was replaced by a new President and CEO.

In February 2002, MDCVU announced its first implant of the Omnicarbon valve in the United States. In March 2003, the company was delisted from NASDAQ.

On April 17, 2003, the company received an allowance for a patent application for its proprietary carbon coating process used to manufacture the Omnicarbon Series 4000 heart valve. Previously manufactured Omnicarbon valves purchased from CarboMedics, Inc. were called Omnicarbon Series 3000. The Omnicarbon heart valves sold to date totaled about 35,000. The company is to be commended for this achievement, one that the Italian biomedical company was never to touch.

On April 19, 2003, Medical CV, Inc. announced reorganization into two divisions. The Heart Valve Division was headed by the new President and CEO who also had the responsibility for international sales of the company's products.

The performance of the Omnicarbon heart valve, as described by the literature reported through 2001, was confirmed as excellent [6–13]. More current unfavorable results were reported in 2016 [14].

In 2016, Kuramisawa and his colleagues reported a 61-year-old woman presented with exertional shortness of breath. She had undergone mitral valve replacement with a bioprosthetic valve in 1985, and this was replaced with an Omnicarbon valve 9 years later.

The 2016 laboratory findings revealed hemolytic anemia. Echocardiography showed an increased mean pressure gradient through the mitral valve and moderate to severe regurgitation around the minor orifice of the tilting disc valve. She underwent a third operation. Pannus—scar tissue formation—was found on the prosthetic valve sewing cuff, but it did not obliterate the orifice.

After the valve was removed, the posterior wall of the left ventricle was seen to be associated with thickened endocardium. A bileaflet valve was implanted. Postoperative echocardiography showed the implanted valve to be functioning well.

Ironically, by the time the Omnicarbon 4000 valve using Medical Incorporated's internally produced Pyrolite Carbon reached the market in 2003, the Medical Carbon Research Institute, founded in 1994, had developed its On-X valve technology that received the CE mark for sale in Europe and FDA approvals for sale in the United States leading to Medical Incorporated's demise.

Shiley's Struggles with Strut Failure

Shifting back to Shiley Laboratories: In 1979, the shape of the Björk-Shiley Pyrolite disc was changed from plano-concave to convexo-concave (to improve hemodynamics, we were told) and the valve relabeled the BSCC valve (Fig. 7.3). This time,

Fig. 7.3 Photograph of the Bjork-Shiley convexo-concave heart valve. Tilting disc is held in place between fixed inlet strut and welded smaller outlet strut. (Reprinted from *Circulation*, Twenty-five-year experience with the Björk-Shiley convexoconcave heart valve, WJ Blot, et al., 2005;111(21) with permission from Wolters Kluwer Health, Inc.)

Medical Products Division's tooling charge was paid for by Shiley Laboratories. While the change helped the hemodynamic performance of the Björk-Shiley valve, the incidence of strut fractures increased. Not surprisingly, Shiley Laboratories' dialogue with the FDA intensified.

In 1980, Shiley Laboratories' Vice President of Operations became the President of Shiley. The strut failures continued until 1986, when the valve was withdrawn from the market by the FDA.

The Shiley experience was sad. The magnitude of the problem had not been recognized until a Special Report appeared much later, entitled *Twenty-Five-Year Experience with the Björk-Shiley Convexo-Concave Heart Valve: A Continuing Clinical Concern*, published on May 31, 2005 [15].

The full text of that report is available online. The summary is comprehensive.

In the same issue, a telling editorial appeared [16], *Could It Happen Again? The Björk-Shiley Convexo-Concave Heart Valve Story* by Eugene H. Blackstone MD. The editorial appears on page 2717 and the full paper on page 2850 (*Circulation*, May 31, 2005).

A statement from the FDA website (Medical Device Recalls, The BI 108 Organ Replacement website, Spring, 2007) included a revealing list of grievances which were numerous and complicated.

The scrutiny of the history of the Björk-Shiley valve was skin-deep. Shiley had misplaced confidence in Delrin as an occluder and on metals for its orifice. Nature taught otherwise: the choice of any metal rather than carbon for long-term use in valves *must* take into account that valves might experience a billion cycles. Consequently, failures of valve struts persisted until the valve's final withdrawal from the market in 1986.

Tellingly, throughout the entire Shiley history, there was not a single failure of the carbon disc occluder.

The folks at Karolinska Hospital in Sweden [17], on a redesign of the Shiley prosthesis, produced an improved device that reduced fatigue fractures and increased orifice area, but it was too little too late. The all-carbon St. Jude Medical bileaflet was soon becoming popular in the 1980s. Even so, Dr. Viking O. Björk continued to pursue his design.

An article titled "The Designer of Faulty Heart Valve Seeks Redemption in New Device," by Barry Meir, appeared in *The New York Times* on April 17, 1990 [18]. The *Times* wrote, "Dr. Björk's story is a tale of a gifted scientist when his prized work went wrong, through no direct fault of his own. Even Dr. Björk's critics agree that he was not responsible for the valve failures."

The Key to the Future

The limit of a valve's useful life approaches a billion heart beats. Blood compatibility is an obvious factor here. With nature's valves, wear and fatigue can be healed, but prosthetic valves aren't capable of self-healing. The frictional wear resistance of carbon rubbing against carbon is less than the wear of carbon caused by metal or the wear of metal caused by carbon [19]. All materials other than carbon are subjected to loss of strength caused by cyclic mechanical stresses; the engineering term for this is fatigue. (Notably, pyrolytic carbon is unique in being immune to failure by fatigue.) The failure of the metal struts in the Björk-Shiley valve is an example of such fatigue [20].

The superiority of carbon compatibility was demonstrated by an increase in thromboresistance (by a factor of one-half) experienced when the Omniscience valve was converted to all-carbon in the Omnicarbon valve.

In 1991, Pergamon Press USA published a book titled *Replacement Cardiac Valves*, edited by Endre Bodnar and Robert Frater [21]. The second chapter by Bokros, JC et al., Durability of mechanical heart valve replacements: past experience and future trends, pages 21 through 48, together with References [22, 23], marks an important time in the development of valve replacements, when a paradigm shift would be forthcoming not only in carbon process but also in valve design. The period began in 1992 with an improvement in carbon deposition technology. This development was significant because it made it possible to design a valve that would preserve natural flow.

The book contained the history of the replacement valves from the first Hufnagel valve through to the development of Pyrolite Carbon valves. The first all-carbon bileaflet valve was yet to come.

The fourth Charles E. Pettinos Award was announced and presented to me (Fig. 7.4) at the 12th Biennial Conference on Carbon in Pittsburgh, Pennsylvania. The Pettinos Lecture was published [24].

Fig. 7.4 The fourth Charles E. Pettinos Award was presented at the 12th Biennial Conference on Carbon in Pittsburgh PA to me in recognition of my "accomplishments in the application of pyrolytic carbon technology to the development of biomedical prosthetic devices". (Used with the permission of Jack Bokros)

References

1. Letter to *Circulation* 76;53:2016.
2. Starek PJK. Chapter 18: heart valve replacement and reconstruction: clinical issues and trends. In: PJK S, editor. Advantages and disadvantages of mechanical valves. Chicago: Year Book Medical Publishers, Inc; 1986. p. 221–35.
3. Callaghan JC, Coles J, Damie A. Six year clinical study of the use of the Omniscience valve prosthesis in 219 patients. J Am Coll Cardiol. 1987;9(1):240–6.
4. Teijeira FJ. Long-term experience with the omniscience cardiac valve. J Heart Valve Dis. 1998;7(5):540–7.
5. Misawa Y, et al. Twenty-two year experience with the omniscience prosthetic heart valve. ASAIO J. 2004;50:606–10.
6. Kazui T, Yamada O, Yamagisi M. Aortic valve replacement with Omniscience and Omnicarbon valves. Ann Thorac Surg. 1991;52:236–44.
7. Misawa Y, Hasegawa T, Kato M. J Thorac Cardiovasc Surg. 1993;150(1):168–72.
8. Weiss P, Jenzen HR, Hoffmann A, et al. The Omnicarbon tilting-disc heart valve prosthesis A clinical and doppler echocardiographic follow-up. J Thorac Cardiovasc Surg. 1993;106(4):599–608.
9. Thevenet A, Albat B. Long term follow-up of 292 patients after valve replacement with Omnicarbon prosthetic valve. J Heart Valve Dis. 1995;4(6):634–9.

10. Abe T, Kamata K, Komatsu K, et al. Ten years' experience of aortic valve replacement with the Omnicarbon valve prosthesis. Ann Thorac Surg. 1996;61(4):1182–7.
11. Iguro Y, Moriyama Y, Yamashita M, et al. Clinical experience of 473 patients with the Omnicarbon prosthetic heart valve. J Heart Valve Dis. 1999;8(6):674–9.
12. Torregrosa S, Gomez-Plana J, Valera FJ, et al. Long-term clinical experience with the Omnicarbon prosthetic valve. Ann Thorac Surg. 1999;68:881–6.
13. Misawa Y, Fuse K, Saito T, et al. Fourteen year experience with the Omnicarbon prosthetic heart valve. ASAIO J. 2001;47(6):677–82.
14. Kurumisawa S, Kaminiski Y, Muraoka A, Misawa Y. J Cardiothorac Surg. 2016;11:40–4.
15. Blot WJ, Ibrahim M, Ivey TD, Acheson DE, Brookmeyer R, Weyman A, Defauw J, Smith JK, Harrison D. Twenty-five-year experience with the Björk-Shiley convexo-concave heart valve: a continuing clinical concern. Circulation. 2005;111:2850–7.
16. Blackstone EH. Could it happen again? The Björk-Shiley convexo-concave heart valve story. Circulation. 2005;111:2717–9.
17. Björk VO, Lindblom D. The Monostrut Björk-Shiley heart valve. J Am Coll Cardiol. 1985;6:11142–8.
18. The Designer of Faulty Heart Valve Seeks Redemption in New Device, Barry Meir. *The New York Times*, April 17, 1990.
19. Bokros JC, Haubold AD, Akins RJ, et al. The durability of mechanical heart valve replacements: past experiences and current trends in. In: Bodnar E, Frater RWM, editors. Replacement cardiac valves. New York: Pergamon Press, Inc; 1991. p. 21–48.
20. Haubold AD, Bokros JC. Chapter 1: carbon in medical devices. In: Williams DF, editor. Biocompatibility of clinical implant materials, CRC series in biocompatibility, vol. 2. Boca Raton: CRC Press, Inc; 1981. p. 3.
21. Book entitled *Replacement cardiac valves*, Book eds by Endre Bodnar and Robert Frater. Pergamon Press USA, copyright 1991.
22. Ibid, Chapter 2, Bokros JC, et al. "The Durability of Mechanical Heart Valve Replacements: Past Experience and Current Trends," Pages 21–48.
23. Ibid, Chapter 15, Schoen FJ. Materials Considerations for Improved Cardiac Valve Prostheses.
24. Bokros JC. Carbon in medical devices (The Pettinos lecture). Carbon. 1977;15:355–71.

Chapter 8
First Bileaflet Valve: St. Jude Medical Inc. Start-Up

Early in 1979, I received a call from an assistant at St. Jude Medical asking about a possible meeting with two executives. They wanted to discuss a potential business arrangement. I was, of course, happy to meet with them, and our meeting was put on the calendar.

Rumors were coming in from friends in the carbon field who thought St. Jude Medical's project had something to do with carbon.

St. Jude Medical was a start-up, founded in the mid-1970s to develop a heart valve replacement. None of the founding team had experience in the development of heart valve replacements but identified an opportunity to enter a rapidly developing medical device field. They were motivated by a valve replacement conceived by an engineer, also inexperienced in the science and technology associated with heart valve replacements in humans.

On the day of the appointment, the two executives were very personable. After pleasantries and sharing with us how much they admired General Atomic's advanced nuclear endeavor, they paused to show me a patent drawing for a replacement heart valve. The Possis patent [1] had already been licensed to St. Jude Medical (Fig. 8.1). The valve was to be manufactured from titanium metal. The marketing executive said that he believed that the new concept would revolutionize the industry. Both were interested in our thoughts on the valve.

St. Jude Medical recognized that the trend of the mechanical heart valve market was increasing for devices that were completely coated with carbon, notably the DeBakey aortic valve (Chap. 5) and the Beall Model 106 valve (Chap. 6).

I held back politely, saying, "Very interesting." After a pause, I told St. Jude Medical the temperatures involved in the deposition process exceeded 1300 °C. Visually, that's white hot—hotter than red, or yellow. I asserted that titanium metal could not sustain the deposition environment. The pivots, perhaps rubies, would also lose viability. The orifice, in two pieces, cylindrical and threaded so the leaflets could be inserted, would be prone to thrombose. The seam between the two pieces would be a site for nucleation of clotting.

J. Bokros, *Heart of Carbon*, https://doi.org/10.1007/978-3-031-17933-4_8

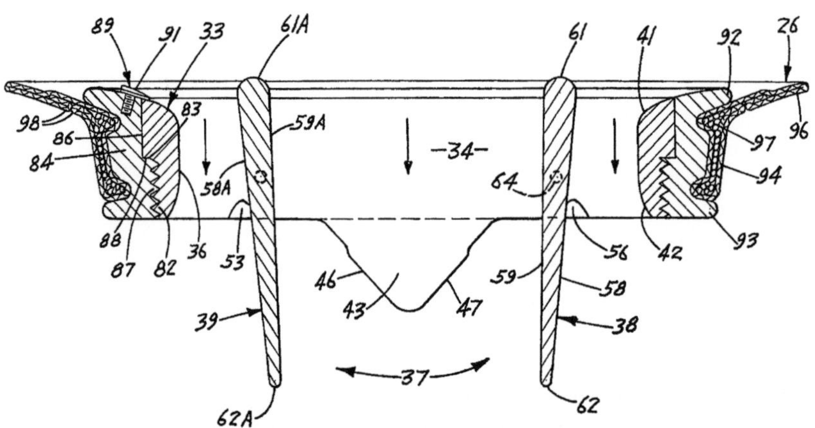

Fig. 8.1 Possis Z., US Patent No 4,078,268, Heart Valve Prosthesis, issued March 14, 1978

The St. Jude Medical executives, inexperienced, were stunned by my assertions, but accepted that the Possis titanium design could not be coated with carbon.

Keeping the meeting cordial, and so as not to offend the two executives, I explained that "We can't make that."

Clearly, I pointed out, heart valve replacements are dramatically different than pacemakers. The executives were seasoned pacemaker people. Pacemakers are comparatively easy to implant, are superficially located in a neat pocket to accept the titanium pacer, and are surrounded by soft tissue (which tolerates the man-made intruder more readily than high-velocity blood would).

I explained to them that blood was finicky and reacted by clotting and rejecting foreign bodies. Carbon has been proven not to react as a foreign body. It is benign in blood and durable enough to last a lifetime—for anyone.

I was wondering how to proceed, not wanting to lose this opportunity. I talked about valve technology, the need for a really good concept, praising the bileaflet design that might be redesigned and made with carbon. I ended up lecturing the two about the design features needed to provide hemodynamics that provided natural flow without clotting and which would last a lifetime.

The ball was now in my court. I was sure that the oil company lawyers (investors) would be leery of making a life-sustaining device out of a material they had commercialized, their fevered imaginations visualizing some such headline as "Oil Company Kills Thousands." That kind of fear paralleled my thinking when Arthur Beall's Model 105 all-carbon valve suffered some failures. The fact was that there had been only half a dozen Model 105 failures—but in this field, one failure is too many.

General Atomic, in past projects, insisted on a strict "hold harmless" agreement for any component supplied and used in a life-sustaining device, to protect against any product liability that might arise if the device should fail in a patient. I advised St. Jude Medical further that all of General Atomic's carbon technology was patented, confidential, and proprietary.

The final understanding was that General Atomic would undertake the St. Jude Medical start-up task at no cost, provided that, should the project achieve its objective, St. Jude Medical would purchase the components exclusively from Medical Products Division.

I was advised by General Atomic's management to avoid liability stemming from the *actual design* of a valve for St. Jude Medical: "Don't patent anything related to their carbon components." However, any technology we developed would remain the confidential, proprietary property of General Atomic. A formal agreement was executed between the companies.

I was aware of St. Jude Medical's reluctance to be tied to a single source. Their people had searched far and wide for other carbon technologies that were not proprietary and confidential. Entering into an agreement with General Atomic was undoubtedly their last resort.

The St. Jude Medical project moved forward. The manufacture of Pyrolite Carbon and the machining of a special form of graphite used for substrates were all unique to this development effort. The thermal expansion coefficients for the graphite had to match that of Pyrolite Carbon.

As we were proceeding, St. Jude Medical persisted in pursuing a license on the carbon technology. St. Jude Medical sent another executive on a fishing expedition. He was a high-ranking officer of St. Jude Medical who continued the aggressive pursuit of a license for the carbon process—as if the success of the St. Jude Medical venture depended on it.

At our first meeting (and memorable it was), the executive arrived in suit and tie, sporting the kind of pipe that professors use as a prop to toy with while pontificating (light pipe briefly, snuff it out, clean the bowl, repeat, put back in pocket, etc.).

He sat back and told me how things must be. St. Jude Medical needed a license. St. Jude Medical needed to be 100% in control. This was an absolute necessity—non-negotiable.

I got the picture and realized that St. Jude Medical needed absolute clarity concerning what General Atomic would and would not do. I considered bending over backward for this important customer.

I told the executive that I understood St. Jude Medical's position completely and suggested that we might be able to persuade my management (the powers that be) to consider setting up, at our own expense, a facility in Minneapolis. This facility would be owned by General Atomic, fully integrated to produce St. Jude Medical's carbon components to specification, and would transfer components to St. Jude Medical at a reasonable price.

I paused to see where he would take this. If he balked at my first pitch, I was prepared to consider letting St. Jude Medical use its own employees to operate the facility. St. Jude Medical might be able to purchase the facility after our patents had expired, at a negotiated price.

The executive expressed interest but had to discuss my proposal with the St. Jude Medical Board. No response was forthcoming.

Robert Akins, Vice President of Operations, proceeded on an OEM basis, using existing validated processes and protocols to produce components to be assembled by St. Jude Medical.

Accommodating Demands for an Inspection

Our fabrication processes were kept secret from St. Jude Medical, and their ignorance of our trade secrets prompted them to express concern about whether we could truly scale our processes to produce valves at commercial volumes. St. Jude Medical finally demanded an inspection to satisfy itself and their investors of Medical Products Division's ability to supply large quantities. They proposed that an engineer that I respected perform the inspection.

I had to discuss St. Jude Medical's proposal with the management team to get the buy-in I needed. The next day I invited the engineer to our facility, but the meeting was strictly confidential.

St. Jude Medical's engineer arrived for the inspection and was escorted by John Sommerfeld, Vice President of Quality Assurance. Sommerfeld was rigid when it came to quality control. In fact in the early days when the Medical Products Division was located in Sorrento Valley, during inspection, if quality control encountered a blemish and could not decide if it was okay, the decision to pass was tossed to me. In one instance, when the blemish came to me, I said "Okay, pass it."

Sommerfeld retorted "If there were consequences, you'd fire me." He rejected the blemished part.

St. Jude Medical's engineer was amazed and happy to see our production facilities were using computer-controlled machines with multiple stations, producing precision graphite substrates and Pyrolite valve components.

Post-inspection, the St. Jude Medical engineer prepared a detailed report, which he shared with us. The report assessed our work to be outstanding and that we knew what we were doing, all with high efficiency. While I was pleased to receive the glowing report, I later regretted that we had shared so much information with St. Jude Medical, even though they had agreed to hold it in the strictest confidence.

Reference

1. Possis Z. US Patent No 4,078,268, Heart Valve Prosthesis, issued March 14, 1978.

Chapter 9
General Atomic Inc. Medical Products Division Sold to Intermedics Inc.

At the same time that we were dealing with St. Jude Medical, General Atomic, controlled by two oil companies, was considering having another large energy company purchase an equity position in General Atomic's HTGR project.

The energy company did not understand what the Medical Products Division was all about and why it hadn't been spun off long ago. They thought Medical Products Division's "dabbling" in medical technology was a distraction from General Atomic's main focus: the energy business.

An announcement was made of the decision to sell the Medical Products Division. This decision to put Medical Products Division up for sale was kept secret from me, so when I heard the startling news from an oil company Executive Vice President, I was both frustrated and angry. It made me feel like a Christmas turkey—without the benefit of prior notice. I had thought that General Atomic would at least discuss the selling of the Division with me prior to making their "for sale" announcement.

The supreme irony was that the energy company never made the anticipated investment. The Medical Products Division sale being put in motion greased the skids. The decision to sell had been made: there'd be no going back.

I followed up the Executive Vice President's announcement with the request that I be given the opportunity to raise money to buy the Medical Products Division. At the time, the annual sales of the Division were about 4 million dollars. An unsolicited investor who had heard the rumor had already approached me and offered 2.5 million dollars, subject to his due diligence.

The Executive Vice President's response to my interest in buying the Division was an insistence that I step aside because with me as a potential buyer, there would be a conflict of interest. He assigned me to a new office in a different building: an empty room with only one telephone resting on the bare floor. No desk or chair.

I decided that my home office would better serve my immediate needs.

Several months went by. The maximum offer I could muster was $5.5 million.

J. Bokros, *Heart of Carbon*, https://doi.org/10.1007/978-3-031-17933-4_9

General Atomic received no offers. The $5.5 million dollar offer was presented by an agent representing a foundation for responsible parenting. I later learned that the "foundation" was really a front for Marshall Kriesel, President of Medical Incorporated, who envisioned cornering the market on medical carbon, thus solving *his* sole source dilemma and in turn inflicting that issue upon our other customers.

General Atomic had not ignored the fact that the Division had, in reality, a monopolistic position covered by complete patent protection. All of our customers had high anxiety about who might buy the division and, in effect, control them through the supply of carbon for medical applications.

All the while, our production of components for the St. Jude Medical valve was on the rise.

An expression of interest in Medical Products Division was put forward by 3M. American Hospital Supply Corporation also expressed interest in acquiring Medical Products Division. Both bids stalled because I wasn't anywhere to be seen. As it turned out, neither company was interested in Medical Products Division without me being on staff. A standoff developed.

Growing anxious about the situation, I decided that my best bet was to withdraw my interest in buying the division and to provide assurances that I would assist in the sale. After 2 weeks of punishing silence, the Executive Vice President finally responded to my idea and agreed to move forward on that basis. He let me know that the oil company would, under no circumstances, ever be intimidated and that he could easily pull the plug and write the division off entirely rather than have someone force his hand.

Shortly thereafter, I received a late night call from Albert Beutel, President and CEO of Intermedics Inc., a new and fast-growing enterprise in the pacemaker business. Beutel told me that he knew I was attempting to buy the Division and wanted to know if it would be okay with me if he made an offer. I told him I had not been successful in raising the required capital, so he should feel free to enter the game. I respected Beutel for the gesture.

3M and American Hospital Supply Corp. presented employment contracts to me and the others that I had identified as my management team, namely, Robert Akins (Operations), Axel Haubold (R&D), and John Sommerfeld (Quality). The oil company Executive Vice President shunned 3M and American Hospital Supply Corp. as being too big to acquire General Atomic's carbon monopoly.

My management team and I decided to include Michael R. Emken as a member of our team. Michael had been instrumental in the development of the carbon process and had earned a promotion.

Beutel offered a contract that was, as you might imagine, entrepreneurial. You can also imagine whose offer we favored: the one coming from Intermedics, Inc. As politely as possible, we declined to sign any of the other bidders' employment agreements. I was careful not to exhibit any discontent, as I did not want to risk irritating the oil company Executive Vice President, and I was looking forward to joining Beutel.

Fig. 9.1 Medical Products Division personnel at the time of the sale to Intermedics, Inc. Four of the management team are on the left: Jack Bokros; Axel Haubold (front row from left to right); Robert Akins; and Michael Emken (back row, second and third from left). (Used with the permission of Jack Bokros)

The most difficult part of the transaction was to get all of the customers to sign off on the deal. Intermedics, Inc. provided assurance that it had no intention to enter into the heart valve field and that the Medical Products Division, which we named CarboMedics Inc., would be separate and autonomous. All parties eventually went along with the proposal. The sale was consummated, effective January 1, 1979. Intermedics ultimately paid $6 million in cash, with some more paid in Intermedics stock.

The Medical Products Division personnel at the time of the sale are shown in Fig. 9.1.

Chapter 10
CarboMedics Inc. Gets Acquainted with Intermedics Inc.

The General Atomic's Medical Products Division was renamed CarboMedics, Inc. with the President of CarboMedics, Inc. reporting directly to the President and CEO of Intermedics, Inc. and CarboMedics, Inc. being set apart and functioning autonomously. Shortly thereafter, CarboMedics, Inc. relocated to Austin, Texas (Chap. 11). Intermedics, Inc. was privately held, and Albert Beutel, President and CEO, was an aggressive entrepreneur.

The employment contract that the five principals of CarboMedics, Inc. signed with Beutel provided "phantom stock" that "should be filed away and not shown to anyone," specifically not to accountants. Beutel certainly had a style of his own, often pushing aside bureaucratic red tape.

He offered 4% of gross sales as an incentive to me to distribute as I saw fit to the five principals. Following "as I saw fit" quite literally, I distributed portions to each and every one of the employees according to my own assessment of the value of their contribution to the company. Later, Beutel sent a note: "I like the way you did that."

In Angleton, Texas, where the main Intermedics, Inc. facility was located, Beutel purchased a gas station to provide free gas to every employee stationed at that facility. Many executives had company cars, which were usually BMWs. At the time, CarboMedics, Inc. was located in La Jolla, California, where none of us received any free gas whatsoever (one obvious downside of being autonomous from Intermedics, Inc.).

Beutel bought all the available space in Angleton, Texas; Intermedics, Inc. pacemakers were very popular. He needed to acquire space fast, as sales tripled in those first years, and he needed the room to grow.

To entice clients to Angleton, Beutel maintained a fleet of aircraft: a large Hawker, a King Air, a Learjet, a smaller Bonanza, and two helicopters. For special events, he had a two-bedroom yacht moored at the Angleton Marina. My first meeting as an employee with Beutel was held on that yacht. The attendees were

© The Author(s), under exclusive license to Springer Nature Switzerland AG 2023

J. Bokros, *Heart of Carbon*, https://doi.org/10.1007/978-3-031-17933-4_10

executives and included Russell C. Chambers, MD, the Medical Director, who sat in a corner. His longish hair and beard gave him a very distinctive appearance.

The meeting was for introductions and pleasantries. Dr. Chambers gave me only a nod of acknowledgement when I was introduced. After the meeting, as we arrived back at the airport and departed the helicopter, Dr. Chambers was directly behind me. He said softly, "If you're so smart, how come you're not rich?" Those were the first words he spoke to me. I was a bit startled and replied, "I hadn't put my mind to it." Perhaps my reply was in line with the theme of Beutel's company.

As I got to know Chambers, I learned his words were often cryptic and mostly jocular. It wasn't long before we became good friends. He often appeared at the CarboMedics, Inc. facility in Austin, putting on airs as if he owned the place. He and I also traveled a lot to conferences, medical schools, distributors, and surgeons—all in the course of conducting business throughout the world.

But Chambers and I also explored mutual investment activities that we anticipated would blossom, opportunities that were not related to Intermedics, Inc. His obsession, in his own words, was to become "filthy rich." I felt that he must be trying to bring me along, to enlist me in the same project. I believe Chambers was what the Harvard Study would call an entrepreneur.

We at CarboMedics, Inc. had no complaints about not getting free gas. We received "phantom stock" instead and were rather content with our autonomy. Intermedics, Inc. was indeed a technology star. Beutel's attributes paralleled those of Dr. de Hoffmann, the founder of General Atomic (which was now a major star only in energy technology, having just spun off its medical technology).

Although autonomous, I was asked to attend Beutel's employee meeting so that I could become acquainted with the key players.

Pat Gordon was a wizard in understanding pacemaker electronics, both theoretically and practically. Jacques Schwartz, head of regulatory affairs, was a pro in the field, making sure that all government regulations were adhered to. Don Hunter, the kingpin of manufacturing, was responsible for Gordon's pacer schematics being miniaturized and sealed in a titanium "can." Steven Anderson, head of marketing, was experienced and innovative in portraying Gordon's electronic magic, along with his sidekick Kelly Atkinson, young and tall and sporting long white hair.

I was impressed by the way this team promoted its products. Their promotional material was accurate and exciting. I imagined that when technical journals arrived on the desks of doctors, the first items they'd flip to were Steve's and Kelly's Intermedics Inc.'s messages.

One of their masterpieces was an advertisement showing Medtronic, Inc.'s pacer as a "hockey puck," black and bulky, compared to a small, gleaming Intermedics, Inc. device exuding the elegance of a da Vinci painting. Beutel's pacemakers, portrayed by Anderson/Atkinson, designed by Gordon, and fashioned by Hunter, had a picture of a rearview mirror labeled "Keeping Medtronic in the Rearview Mirror."

Dick Martin played his role well as a professor charged with educating cardiologists to understand electronic fundamentals. He was also responsible for the instructional documentation, which was as clear and concise as a flight manual.

Sid Barbanel was head of sales, very vocal and colorful, aggressively forceful and savvy enough that Beutel knew, by the hour, the firm's sales numbers. He spelled Pyrolite *pirolight*—a mistake I didn't point out to him.

Steve Anderson operated in the midst of Beutel's Intermedics, Inc. and understood its culture very well. He mentioned that Albert Beutel was the grandson of the oilman who built the huge oil refining complex located near Angleton. Beutel, recognizing that his grandfather's reputation could be an advantage, founded Intermedics, Inc. there. Anderson added that when Beutel approached the local bank, he carried a large briefcase filled with Krugerrands in the hope of making an impression.

Beutel was like one of the boys but was admired like a king. Pat Gordon jokingly told me about the time Beutel bought a mobile home, a top of the line model, like his aircraft, that was as big as a Greyhound bus. When it was delivered, he jumped in and declared, "I can drive this thing!" He revved up the engine and it stalled. Beutel muttered some vulgarity. But, since he had an audience, he got it running and, like a pro, slammed it into gear—the wrong one. The transmission, according to Gordon, fell out onto the ground, gears and all.

Gordon, breaking up in laughter, described the bus being towed back to the garage with Beutel still at the wheel. The laughter was assuredly nuanced so as not to upset Beutel. Fred de Hoffmann would never have got himself in such a pickle. De Hoffmann was charismatic but always maintained a dignified image.

About a month after the sale of Medical Products Division to Intermedics, Inc. Beutel, late at night as was his custom, called me:

> "Put some engineers out of sight and make a valve for us. I'll call you later." Brief and to the point, he hung up before I could reply. I interpreted Beutel to mean "in a secure proprietary location."

I was stunned because, in the negotiation of the sale of Medical Products Division to Intermedics, Inc., Intermedics had affirmed that it had no intention of going into the valve business directly. Beutel was required to have "no intention" to go into the valve business because, without that provision, there would be a restraint of trade and conflict with existing antitrust law.

I was slow to react to Beutel's orders. I had supply contracts with people at Surgitool, Medical Inc., Shiley Laboratories, and St. Jude Medical who I considered friends. To me, Beutel's instructions would be a double-cross, clear and simple. I lost sleep over that phone call.

In March of 1979, when I was still living in Alpine, California (it being only 3 months after the acquisition), I received a call from Jacques Schwartz informing me that Albert Beutel had been killed in a helicopter crash, along with the pilot. Jacques told me that most everything was now on hold.

Anderson vividly related the notable incident: "Beutel had a difference of opinion with the CFO. Arriving on site with a handgun, Beutel set out to fire the CFO." Anderson further relates, "that Beutel called a telephone board meeting and fired the CFO. Beutel was so mad that he charged out of his office and got into his helicopter to go home. On the way home, the helicopter crashed, killing Beutel and the pilot."

As could be expected, the death of Beutel, the President and CEO, was a blow precipitating much anxiety and unusual behavior at the executive level of

Intermedics, Inc. In fact, I was told that Barbanel had jumped up on a chair to declare that *he* could run Intermedics, Inc. Beutel's unique entrepreneurial drive made it unlikely that the board would ever find another Albert Beutel. I was concerned because every one of the executives thought they could take charge. The thought of having to report to any one of them was enough to make me shudder.

I breathed a sigh of relief when the Board announced that G. Russell Chambers (Fig. 10.1) was selected as the new President and CEO. GR (the nickname his friends used for him) was a seasoned businessman and the father of Russell C. Chambers.

The closing of ranks behind Chambers was not a simple matter.

GR's close friend was Wayne Rollins, a mature entrepreneur who had recognized Beutel as a good bet. The story goes that a tripling in sales required high cash flow. Rollins was always there, so once when Beutel called him, short of cash, Rollins instructed him to "fly to Houston and pick it up at the Warwick Hotel." Beutel flew his Bonanza and got a $900,000 check.

Once Beutel landed back in Angleton, he discovered that the check had gone missing. It was later found in a flower bed on Galveston Island.

From the acquisition, I had gained considerable respect for GR's style, which was steady, never precipitous, always shrewdly calculating, but with a friendly demeanor. I found him easy to communicate with, and we became friends.

When I told GR about Beutel's order to make a valve, he froze, blinked, and pursed his lips as he chewed on that revelation. "Let's think about that. Hold off. I'll get back." He never did get back to me. His silence was welcome, and obviously necessary, because we both knew Beutel's proposal was unethical.

Fig. 10.1 Russell Chambers, President and CEO of Intermedics, Inc. (Image provided with courtesy of Boston Scientific. ©2019 Boston Scientific Corporation or its affiliates. All rights reserved)

Chapter 11
CarboMedics Inc. Moves to Austin Texas

In 1979, our lease in Sorrento Valley would expire. The management decided that California taxes were too high, and on the rise, so we needed to relocate.

We first concentrated on a state that had no state income taxes. Washington State had no state income tax, and because of its location, the cost of electric power was low. We then looked at Florida, but finally decided that San Antonio and Austin were better options.

Two factors helped settle the final decision.

First, the front cover of *Money* magazine had listed the cities in the United States most attractive financially to small businesses and start-ups. The city at the top of the list for overall desirability was Austin, Texas.

The second factor was financial in nature. When Akins and I had chatted with GR, the new President and CEO of Intermedics, Inc., about the move, he sat back to think. "If it were me, I'd go to Austin!" he told us in his distinctive Louisiana drawl. GR let that sink in with us and then added, "If there's trouble in Texas, all the money goes to Austin."

It was a prescient insight. At this writing, I recall a similar statement years later when CarboMedics, Inc. was considering placing a valve assembly plant in China. The people from Shanghai told us, "All our money goes to Beijing!"

Losing no time, Bob Akins (Fig. 11.1) traveled to Austin to research and explore the possibilities. It would be his call to decide on a site. Bob located a builder that had first "dibs" (as they say in New York City) on about 20 acres located at the juncture of the main throughways.

Without delay, Bob and I traveled to Austin to meet Cliff Woerner, an established developer, who bought the site specifically to erect buildings for companies like ours (to build a retirement fund for himself, he told us).

Cliff was a Texan. A Texas Ranger lever-action carbine hung over his mantle. When I mentioned that I wanted to have a Spanish style house with a tile roof, he retorted, "You'd best build in San Antonio." When I got around to building a house (in Austin, as it turned out), it had a cement-shingle roof, which was actually cheaper and easier to install than Spanish tile.

J. Bokros, *Heart of Carbon*, https://doi.org/10.1007/978-3-031-17933-4_11

Fig. 11.1 Left to right: Robert J. Akins, Vice President of Operations of CarboMedics, Inc.; Carol Petrie (Dr. Chambers fiancée); Jack Bokros; Russell C. Chambers MD; and Charles Griffin, a CarboMedics, Inc. bioengineer. (Used with the permission of Jack Bokros)

The New Facility in Austin

Bob Akins took charge of the project. As the head of operations, he had to be happy with it. The first building was 50,000 square feet, with room on the site for three more buildings of about the same size.

By the summer of 1981, the first building was finished. Akins leased a temporary building on Beatrice Cove, with no air conditioning, to house the pilot production until the permanent facility was completed. Because of the Texas heat, it was a monumental task. I really appreciated that I had originally picked Akins from the best at General Atomic. He was a jewel!

Akins was responsible for all aspects of a manufacturing facility designed to produce carbon heart valve components manufactured to rigid specification. He was responsible for hiring qualified persons capable of performing according to a variety of skill sets needed to function efficiently in each and every production step.

I'll give an example. The St. Jude Medical prosthesis has been arbitrarily selected to describe the production of its carbon components.

St. Jude components come in six sizes. For each size, there are an orifice and two leaflets. At the first step, substrates were machined from an isotropic polycrystalline graphite purchased from a specialty supplier. Akins wryly observed that the making and grinding of thin graphite substrates was "like machining eggshells."

The next step was to use a proprietary processing furnace to coat the substrate with a carbon material. It is the vertical fluidized bed apparatus described in Fig. 5.2. The equipment was capable of depositing carbon to specified coating thickness and properties.

Each lot of carbon components was finished using grinding and vibratory polishing procedures. The components were contained in a capsule filled with a fluid polishing compound. Following a programmed shaking sequence, the components emerged from the capsule with a surface that conformed to the surface smoothness standards.

Of course, all St. Jude Medical valve components conformed to St. Jude Medical specifications.

Finally, in some cases, a batch of leaflets and orifices were cleaned by outgassing in a "hard" vacuum at 1000 °C for a specific period of time. The vacuum is called "hard" because its pressure approaches the atmospheric pressure of outer space, which is 10^{-8} atmospheric pressure.

All this effort was accomplished by Akins's team of expertly hired and trained persons to carry out the task selected in this example. The details of every aspect of a variety of proprietary procedures (patented or trade secret) were not disclosed to St. Jude.

Indeed, Akins was a jewel, perhaps a diamond, with each of the team members ranked by carets according to their specific capabilities. In fact, our early companies provided incentive shares of company stock, valued at the level of their skill sets (value to the company), to each and every employee.

When GR came in to inspect the new building, he asked, "Why such a wide hall—so much to heat and cool—and why such big offices?"

I answered that the secretaries would be in the hall and added, "My files are all nearly filling the room. The size was because Bob's office was big."

GR frowned. In response, I told him, "Bob thought my office needed to be bigger than his." At this, GR acquiesced: "Okay, Bob deserves it."

We had originally budgeted $250,000 for the move. By the time the move had been completed, the cost had grown to ten times that! When I relayed the overrun—a big one—GR's response was, "Why don't you lose that first budget." He said this with a smile because revenues were rolling in, most stemming from our work with St. Jude Medical. GR's statement reminded me of the phantom stock about which Beutel had pointedly warned us, "Don't tell the accountants!"

New Facility, New Problems

Once production was running smoothly, a problem arose.

In processing polished critical valve components, each one is examined using precision optical equipment to assure that the surfaces are smooth and without flaws (such as pits, scratches, cracks, or other visible defects).

John Sommerfeld (Fig. 11.2) was responsible for product inspections, assuring that every component shipped to St. Jude Medical met their specifications and was free of defects.

The CarboMedics, Inc. inspectors knew a flaw could mean life or death for the implantee, so every part needed to be absolutely free of defects. Each inspector was

Fig. 11.2 Center front: John D. Sommerfeld, Vice President of Quality Assurance of CarboMedics, Inc. Left to right: Tom Waits; Jack Bokros; Randy Henkes; and Michael Emken. (Used with the permission of Jack Bokros)

required to sign their inspection report, making them personally responsible for every component they inspected.

Early in 1981, one of the inspectors questioned a "mark" that she could see—she was not sure if it was a crack or a surface blemish. Other opinions were solicited. The interpretations offered were many and frustratingly inconsistent. As an ad hoc committee, they concluded that cracks might be present but not visually detectable.

All work was stopped. Following procedures, Sommerfeld contacted St. Jude Medical and described the event. He asserted that he was prepared to travel to St. Jude Medical to conduct an inspection of their inventory. Of course, St. Jude Medical was upset, as all shipments were stopped and inventories had to be quarantined until the "event" could be resolved.

It was clear from the variety of interpretations of the anomalous mark that a resolution was urgently needed. All agreed that an appropriate proof test was the answer.

Product liability litigators often try to insert the term "latent defect"—and to put the phrase "free of latent defects"—into supply contracts. A latent defect is a defect that can't be defined: it's a vague catch-all term. The proper defense against that tactic is to demand that the purchaser provide a specification (including configurational drawings and limiting tolerances) for latent defects. Latent defects are never allowed in any sense in a supply agreement.

A well-designed proof test would resolve the question of insuring that the "mark," whatever it was—visible or not—would not affect the mechanical integrity of the valve in vivo. The proof test applied stress levels substantially higher than those expected in an implanted valve for the lifespan of any patient.

The approach was straightforward in spite of the time constraint. A proof testing procedure was put in place. We were back online with components available before the end of 1981. The proof test specifics were treated as a trade secret. The methodology was released to clients only after signing a confidentiality agreement and thus was contractually protected as a trade secret.

If this situation had happened 15 years later, the work of Professor George Sines at UCLA (an expert in engineering, in material science, and, specifically, in fracture mechanics) would have had the answer in hand.

I met Professor Sines at a Carbon Conference after presenting the Pettinos Award Lecture in Pittsburgh. He had a close relative who received a DeBakey aortic valve. As a material scientist, he was intrigued by the properties of the isotropic carbon I was talking about. It was normal to assign the task of understanding the fracture mechanics of isotropic pyrolytic carbon to a student, and Sines's graduate student, Ling Ma, was assigned the task as a thesis topic in pursuit of his PhD.

Ling Ma's near-decade-long effort was exemplary. He demonstrated that LTI pyrolytic carbons possess a remarkable and unique set of mechanical properties, making them virtually immune to fracture due to fatigue, even in the presence of cracks substantially larger than "marks" or hidden latent defects. Ma's conclusions continue to reinforce our confidence in carbon heart valve replacements. A summary of Ma's thesis is provided in the Appendix [1].

Reference

1. Ma Ling. Studies on pyrolytic carbon for biomedical applications. A dissertation submitted in partial satisfaction of the requirements for the degree Doctor of Philosophy in Materials Science and Engineering. Approved Aly H. Shabaik, Jenn Ming Yang, John M. Christie and George Sines, Committee Chair. University of California Los Angeles; 1997.

Chapter 12
The Hemex Inc. Venture

Hemex, Inc. was formed in 1980. Its president, Professor Jerome Klawitter, was an old friend. We met in the 1960s at the first Clemson University Symposium organized by Samuel F. Hulbert, Head of Interdisciplinary Studies at Clemson. Klawitter was a graduate student under Hulbert, pursuing a PhD degree in bioengineering. In 1970, Klawitter became the first person to receive a PhD in Bioengineering from Clemson. When Hulbert moved to Tulane University as Dean of Engineering, Klawitter followed shortly thereafter as an Associate Professor on a tenure track.

Intrigued by the fact that isotropic carbon's elastic properties matched those of bone, Klawitter focused on developing carbon dental implants and finger joint replacements testing his implants in animals. Histological studies of the bone/carbon interface functioning under load demonstrated that cancellous tissue grew directly into the microporosity at the surface of carbon, with no evidence of a foreign body reaction (Fig. 12.1).

Steve Cook, a graduate student under Klawitter, demonstrated the remarkable differences in the compatibility of carbon with cartilage. Compared with metal and ceramics over the long term, carbon was the only material that did not degrade cartilage [1–3].

After the Medical Products Division of General Atomic was purchased by Intermedics, Inc., Dr. Chambers, the Medical Director, focused on CarboMedics, Inc. He was especially interested in the studies of Klawitter and his colleagues at Tulane for non-cardiac applications. Klawitter had two decades of experience working with General Atomic and CarboMedics, Inc., focusing on medical devices implanted in bone.

Thinking outside the box, Klawitter took on the challenge of replacing the cardiac valve with carbon and took notice of the first bileaflet St. Jude Medical valve. Protecting his idea with a confidentiality agreement, he revealed his design to Dr. Chambers (Fig. 12.2).

Like Beutel, Chambers's entrepreneurial instincts identified an opportunity. He sought the advice of an attorney, which led to the formation of a research and development (R&D) partnership to pursue Klawitter's ideas. Dr. Chambers obtained

© The Author(s), under exclusive license to Springer Nature Switzerland AG 2023

J. Bokros, *Heart of Carbon*, https://doi.org/10.1007/978-3-031-17933-4_12

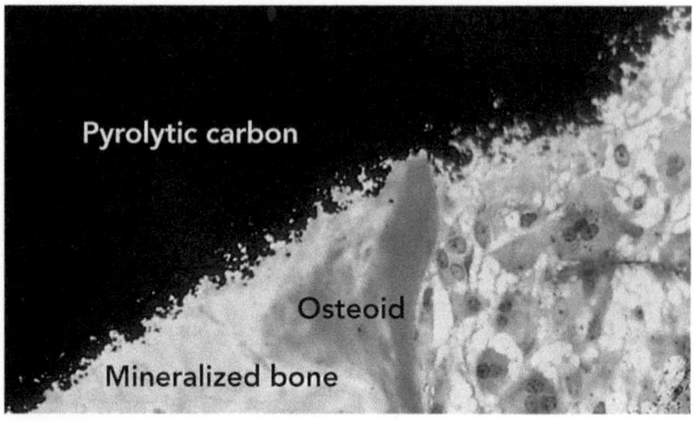

Fig. 12.1 Interface of unpolished porous surface of Pyrolite Carbon-coated dental implant reveals the absence of foreign body reaction. Histological section prepared by Klawitter. (Reproduced with the permission of Jerome Klawitter)

Fig. 12.2 Professor Jerry Klawitter PhD and Russell C Chambers MD. (Reproduced with the permission of Jerome Klawitter)

permission from the Intermedics, Inc. Board to pursue the partnership by using qualified investors to fund development of a valve replacement using carbon components coming from CarboMedics, Inc.

Initially, Klawitter was fully involved with projects at Tulane, but as the partnership developed, he agreed to lead the effort and to eventually receive a royalty should the valve become viable. Looming in the background was the awareness that giving up tenure was not something to be taken lightly.

The Dawn of Hemex, Inc.

Klawitter's idea flourished. A patent application was filed (Fig. 12.3). As the reality emerged, Klawitter's enthusiasm grew. His valve was labeled the Duromedics valve. He gave up his tenure and became the president of a new entity called Hemex, Inc., funded by an R&D partnership of qualified investors.

Hemex was housed in a facility located in New Orleans, close to Tulane University. When animal studies were completed at Stanford University, Klawitter engaged Cliff Woerner to build a facility next to CarboMedics, Inc. on Woerner's development site in Texas, one that Klawitter leased. It was certainly convenient, but St. Jude Medical and other customers of ours were uneasy with a competitor in such close proximity to CarboMedics, Inc.

Heart valve prostheses must undergo 600 million cycles of fatigue testing to insure they are sufficiently durable for long-term clinical use. In order to conduct a 600 million cycle test in a reasonable period of time, fatigue testing is conducted at an accelerated rate of approximately 1000 beats per minute. Even at this high cycle rate, a 600 million cycle test takes more than 1 year to complete.

Klawitter brought in John Ely as VP of Regulatory Affairs for Hemex. Ely, a former bioengineering student of Klawitter's at Clemson, had 7 years' experience at the US FDA.

The key design features of the Hemex Duromedics orifice were as follows:

1. A non-symmetrical curved slot that allows translation as well as rotation (two degrees of freedom) to reduce stagnation of blood.

United States Patent   **4,328,592**

Klawitter   **May 11, 1982**

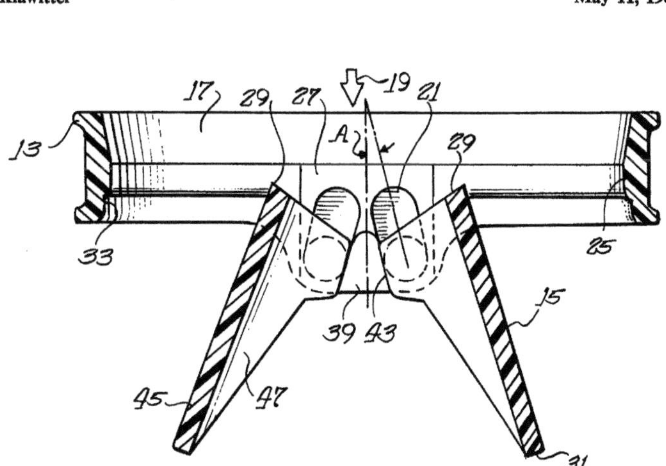

Fig. 12.3 Klawitter JJ. US Patent No 4,328,592, issued May 11, 1982

2. A seating lip within the valve housing, designed so that on closing, forces are transferred from the leaflets to the seating lip, thus "unloading" the hinge apparatus to provide a clear path for purging the socket.

In order to manufacture the Hemex Duromedics orifice, the process for producing Pyrolite Carbon valve orifices had to be modified. The modification, described in the European Patent [4], is called the "mandrel process" (Fig. 12.4).

In this process, all features that occur on the inside surface of the finished orifice are machined in a reverse perspective onto a graphite substrate, which we called a mandrel, which is coated with a thick coating of Pyrolite Carbon. After the deposition, the substrate or mandrel is removed by grit blasting, leaving a freestanding, solid Pyrolite Carbon orifice.

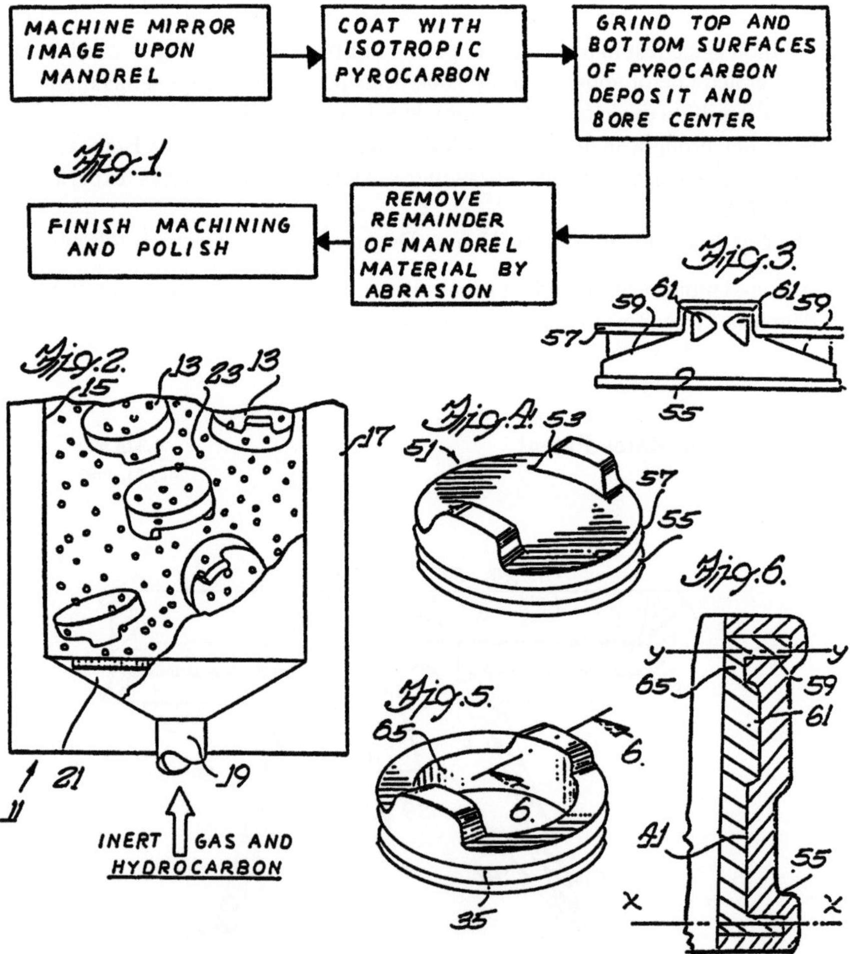

Fig. 12.4 Mandrel process

Michael Emken prepared mandrel substrates for the St. Jude Medical valve as a trial. After coating, he removed the mandrel, leaving a perfect solid carbon St. Jude Medical orifice.

A solid Pyrolite Carbon Duromedics orifice that did not have an internal graphite substrate was accomplished by the following process:

1. Machining a graphite mandrel with the outer surface configured as an exact negative of the internal geometry of the orifice.
2. Coating the mandrel with a thick layer of Pyrolite Carbon.
3. Grinding the outside surface to final dimensions.
4. Removing the graphite mandrel material by differential grit blasting leaving a solid Pyrolite Carbon orifice with precise internal features ready for final polishing.

The process eliminates grinding the inside surface of the orifice, along with the associated grinding defects.

Pyrolite Carbon's low elastic modulus makes it possible to deform the carbon orifice for leaflet insertion. The leaflets are retained in the orifice by spherical protrusions that engage mating sockets in the orifice inner wall. Valves are assembled by mechanically deforming the orifice so that the leaflets could be placed into position; slowly releasing the stretching force allows the orifice to capture the leaflets in position. Clearances between the leaflets and orifice are controlled to maintain proper function.

Since the orifice is capable of being deformed to insert the leaflets, it can also be deformed during function and cause binding or loss of the leaflets. To eliminate this possibility, a cobalt-chromium-molybdenum alloy metal ring is installed around the outside of the valve orifice after the leaflets have been inserted. The high elastic modulus of the metal ring stiffens the assembly, thus assuring free leaflet motion. There had, in fact, been reports of the loss of leaflet(s) of the St. Jude Medical valve during handling and in patients.

The translation of leaflet protrusions within elongated sockets generates a pumping action that flushes the hinge mechanism during each cardiac cycle, thus eliminating stagnation. A seating lip circularly positioned on the orifice wall is situated so that, upon closure, the leaflet balls rest free and clear of the socket, providing for a complete purge of the socket and a transference of the closing impact to the seat.

In vivo animal tests were conducted at Stanford University by Phil Oyer, MD, and his colleagues. Clinical grade Duromedics valves were implanted in 21 calves.

When the animal tests were completed, the results, together with all the test data, were assembled into an application for an IDE to carry out clinical trials in human subjects. The IDE was approved and trials were carried out in Europe. Pre-market approval was received in the United States in 1986.

In 1986, after about 7000 valves were implanted, Baxter (Edwards CVS Division) purchased Hemex, Inc. even though there had been a low incidence of cavitation damage. Until this occurrence in the Duromedics valve, it was thought that cavitation observed in accelerated testing was an artifact of the test and not close to the physiological condition in actual patients.

After the purchase, there were clinical incidents, though infrequent, of cavitation, causing industry-wide concern and triggering scientific investigations into the cavitation mechanism.

Although clinical experience had been good, in 1988, after reports of 12 valve fractures out of 20,000 valves implanted, Edwards terminated the distribution of the Duromedics valve.

After 10 years of follow-up, the Edwards Duromedics valve reported thromboembolic complications at 0.96% per patient year and anticoagulation-related hemorrhage at 1.34% per patient year. Valve failure was 1.54% per patient year. All patients functionally improved (89% in New York Heart Association Classes I and II). The complete reporting is in Reference [5].

The cavitation-related failure of the Duromedics prosthesis led to studies of cavitation erosion and the effects of cavitation on cellular elements of blood [6, 7]. Early reports of cavitation were described and discussed (Fig. 7.1).

Reaching beyond Carbon Heart Valve Replacement

After the sale to Baxter, Klawitter focused on developing carbon orthopedic prosthetic devices utilizing carbon components produced by CarboMedics, Inc. according to his specifications. In the late 1980s, he founded The Little Big Toe Company, Inc. to develop a pyrolytic carbon replacement for the great toe. After commercializing a variety of great toe prostheses, Klawitter founded Ascension Orthopedics, Inc. to expand carbon orthopedic joint replacement prostheses for the extremities, e.g., for ten bones of the hand, wrist, elbow, and shoulder.

In 2011, Ascension Orthopedics was acquired by Integra Life Sciences, Inc. Klawitter at 77 years continues his research on the development of carbon shoulder replacements [3].

During his involvement with Hemex's heart valve replacement and extending through the commercialization of orthopedic replacements, Klawitter found time to work on motor vehicles. Sam Hulbert told me that during the Clemson days, he had to drag Jerry out from under an old car he was restoring. Recognizing Klawitter's talents, Sam thought he needed at least a PhD in Ceramics.

Ultimately, Klawitter turned his attention to the designing of high-speed motorcycles. He set a land speed record of 218 miles per hour, and his latest model achieved 265 miles per hour at the Bonneville Salt Flats. The driver of the motorcycles, Mark Seeley, a master experimental mechanic, worked at CarboMedics, Inc., Medical Carbon Research Institute, OnXLTI, and now CryoLife (now Artivion, Inc.). Klawitter assured me that Seeley's well-being is supported by comprehensive insurance.

References

1. Cook SD, Thomas KA, Kester MA. Wear characteristics of the canine acetabulum against different femoral prostheses. J Bone Joint Surg. 1989;71-B:189–97.
2. Stanley J, Klawitter JJ, More R. In: Revell P, editor. Replacing joints with pyrolytic carbon in book, joint replacement technology. Woodhead publishing; 2008. Cambridge, England and CRC Press. Boca Raton, Boston, New York, Washington DC.
3. Klawitter JJ, Patton J, More R, Peter N, Podnos E, Ross M. In vitro comparison of wear characteristics of pyrocarbon and metal on bone: shoulder hemiarthroplasty. Shoulder and Elbow. 2018;1:12.
4. Bokros JC, Emken MR, Akins RJ, et al. European Patent Specification 0055406, filed 071281 and issued 270385; n.d.
5. Podesser BK, Khuenl-Brady G, Eigenbauer E, et al. Long-term results of heart valve replacement with the Edwards Duromedics bileaflet prosthesis: a prospective ten-year clinical follow-up. J Thor Cardiovasc Surg. 1998;115(5):1121–9.
6. Johansen P. Mechanical heart valve cavitation. Expert Rev Med Device. 2004;1(1):95–104.
7. Andersen TS, Johnasen P, Christensen DO, et al. Interoperative and postoperative evaluation of cavitation in mechanical heart valve patient. Ann Thorac Surg. 2006;81:34–41.

References

Chapter 13
The St. Jude Medical Inc. Litigation

In 1983, an investor visited St. Jude Medical at their Minneapolis facility to discuss the possibility of an investment with their President. He was warmly greeted.

The President was surprisingly frank. He related that St. Jude Medical had a program in full swing to make its own carbon that would increase St. Jude Medical's margin, strengthen its balance sheet, and increase its value. He asserted that the program was going smoothly and that he projected an optimistic view that the completed project would vertically integrate St. Jude Medical and reduce costs. He added that it would be ready to go in short order.

St. Jude Medical's President was baring his soul to a stranger without first requiring the execution of an appropriate confidentiality agreement. This was no ordinary stranger. He was a substantial investor in Intermedics, Inc., and CarboMedics, Inc. was one of its subsidiaries.

St. Jude Medical had a building to house the project, complete with windows that were painted black, which we surmised was to conceal the infringement of CarboMedics, Inc.'s patented technologies.

The day after his visit, I received that investor's comments alerting us of St. Jude Medical's activities.

Intermedics, Inc. set its security folks in motion. A plan was devised to collect the trash that St. Jude Medical was setting out for collection. The yield was rewarding. There was correspondence between St. Jude Medical and our competitor in Italy, who had been solicited to help boost St. Jude Medical's efforts to advance their own Pyrolite Carbon production facility. A computer tape found in the trash turned out to be that of a confidential CarboMedics, Inc. engineering drawing.

The stream of trash revealed that the copycat project was floundering. St. Jude Medical personnel were proving to be naive with respect to carbon processing technology. St. Jude Medical had erroneously surmised that all that was needed was to hire some engineers. The fact is that the CarboMedics, Inc. technology was novel, unique, and securely guarded—engineers with experience in the field simply did not exist in the outside world.

J. Bokros, *Heart of Carbon*, https://doi.org/10.1007/978-3-031-17933-4_13

To gather proprietary information, St. Jude Medical would give notice to CarboMedics, Inc. that the FDA required an inspection of its facilities. When the inspectors returned to St. Jude Medical, they were debriefed by their attorneys about their observations during their visit to CarboMedics.

Intermedics, Inc. President and CEO, GR Chambers, knew that the law did not require one to supply products to an infringer of one's patents, especially when the theft of trade secrets and confidential proprietary information might be involved. GR ordered a shutdown of all shipments of carbon heart valve components to St. Jude Medical and initiated a lawsuit with allegations of patent infringement and theft of trade secrets in Austin, Texas. St. Jude Medical counterpunched with allegations of breach of contract.

In the discussion between the CarboMedics, Inc. attorneys, one attorney, Julius Tabin, took a contrary tack. He said simply, "Why not just let the contract expire?" I listened, but GR had the final say. Julius's comment fell on deaf ears. At an earlier time, Julius had advised: "Jack, don't be precipitous." In this current instance, I should have been precipitous and advised that GR not sue as the only counterclaim that St. Jude advanced was CarboMedics, Inc.'s breach of contract.

Attorney Julius Tabin was once a physicist who had worked with Fred de Hoffman on the Manhattan Project. He told me that after the first test firing of a nuclear device at White Sands, New Mexico, ground zero needed to be investigated. Julius was given a helmet and driven out to ground zero in a tank. One of the speculations was that ground zero might be like talcum powder, and the tank could sink in it, like a stone into quicksand.

Dirty Tricks That Backfired

As pressure mounted to acquire production capability, and with their supply from CarboMedics, Inc. cut off, St. Jude resorted to covert activities. A St. Jude Medical employee called a CarboMedics, Inc. employee (who had previously been an employee of St. Jude Medical) and asked him if he would furnish some information that had evaded their "FDA-mandated" inspections.

Our employee promptly advised me of the ruse. Our security officer immediately notified the FBI. The next phone call between the St. Jude Medical employee and the CarboMedics employee was recorded by the FBI. A list of confidential technical information was transmitted, along with a statement that the CarboMedics employee would perhaps receive a new car.

St. Jude Medical had become convinced that CarboMedics, Inc. had a mole inside the company because CarboMedics, Inc. simply knew too much. When I was asked in a deposition to reveal the identity of the mole, our attorney pondered the question and replied, "Okay, provided that the identity of our source be protected." Our attorney demanded that the magistrate obtain an appropriate order from the Court. The order was granted.

Now properly shielded, I revealed our mole's identity: "We picked up St. Jude Medical's trash."

In keeping with the request for *preserving our source*, St. Jude Medical was required to obtain a secure warehouse and store all their trash there! St. Jude Medical subsequently purchased a number of shredders so if we inspected the saved trash, it would be hard to decipher: our "mole" had been shredded.

During this time period, I had seen an employee of our Italian competitor, the head of their heart valve replacement business, at a conference. I walked up and greeted him with the question, "How are you doing with making St. Jude Medical's valve components?" Off his guard, he answered without hesitation, "We can make the leaflets, but we have yet to make an orifice."

During the court proceeding, St. Jude Medical petitioned the court for a change of venue from Austin to Minneapolis, which was granted. This was surprising because all of the theft of confidential CarboMedics, Inc. secrets took place in Austin. Even the FBI recording of an attempted payoff for confidential information had been made in Austin.

One explanation for the judge's decision to grant a change of venue was that CarboMedics's attorney was a Boston attorney. When venue issues came up, our attorney thought the change of venue was a clear tilt in CarboMedics, Inc.'s favor, so he asked his assisting attorney, the one with a heavy Boston accent, to take care of the issue.

The speculation was that the lady lawyer's Boston accent was not compatible with the judge's Texas boots. In any case, we now found ourselves in Minneapolis: St. Jude Medical's territory.

There was considerable pressure by the court to settle the dispute. On one occasion, the magistrate gestured toward the jury seating area and said, "You will be judged by a jury made up of square-heads [a term used locally to mean less than bright] who wouldn't know the difference between a heart valve and a can of Campbell's soup."

As legal costs mounted (reportedly topping seven million dollars), the parties reached a settlement agreement.

CarboMedics, Inc. agreed to supply St. Jude Medical components at a substantially increased price, and they could manufacture them once they had a viable production methodology, but their in-house output could not exceed 10% of the total number used. Escalation of production beyond the 10% stipulation could occur only after our patents had expired.

For a time after the settlement and until the patents had expired, CarboMedics, Inc. agreed to produce valves in such volumes as St. Jude Medical needed. I estimate that all CarboMedics, Inc. deliveries for valve sets (three components per set) exceeded 400,000 valves. On October 17, 1985, the one millionth component was produced by CarboMedics, Inc. There was no obligation to teach St. Jude Medical how to produce carbon. The Italian biomedical company was still unable to produce the St. Jude Medical orifice.

Some years later, after the St. Jude Medical trial had been settled, I saw that same talkative Italian biomedical company engineer and engaged him in friendly banter.

(I already knew that the Italian biomedical company was working on a bileaflet valve of its own.) He told me that they were still having trouble with their own new bileaflet orifice and were resorting to machining it out of titanium metal. Copying CarboMedics, Inc.'s Biolite® carbon technology, they coated the titanium orifice with a thin vapor-deposited carbon coating deposited at low temperatures that was vulnerable to flaking off. It is interesting to note that the wear rate of carbon bearing against titanium is well known to be an order of magnitude greater than carbon bearing against carbon. For reasons unknown, the Italian biomedical company's design team discounted these decisive factors [1].

As a result, I made a reasonable proposal to the Italian biomedical company's employee, "We could make a proper all-Pyrolite Carbon orifice at a reasonable cost and would be willing to keep the supply contract secret, so that you could sell it as your own." Not tempted, the conversation ended.

Reference

1. Haubold AD, Bokros JC. Carbon in medical devices, in biocompatibility of clinical implant materials. In: Williams DF, editor. , vol. 2. Boca Raton, Florida., Chapter 1, 3: CRC Series in Biocompatibility, CRC Press, Inc.; 1981.

Chapter 14
Broadening the Horizon

When isotropic carbon was discovered in 1963, patent applications were filed, broadly claiming use in medical prostheses. Acquisition of the Medical Products Division of General Atomic by Intermedics, Inc. entailed the acquisition of a patent-protected monopoly through this subsidiary. At the time of the acquisition of Medical Products Division, the President and CEO of Intermedics, Inc. signed a statement designating that Intermedics, Inc. had no intention of entering the artificial heart valve business.

The St. Jude Medical litigation changed our perspective on this restriction.

Before the litigation with St. Jude Medical, St. Jude Medical's growth rate was high. But it lost market share due to the litigation, because their hidden project to produce their own carbon components failed to produce anything. Shiley and other customers of CarboMedics, Inc. moved in to fill the gap. Notably, in 1 year alone, 40,000 Pyrolite occluders were delivered to Shiley. After the litigation, CarboMedics, Inc. did not stop the production of the St. Jude Medical components. Instead, it built up inventory. In fact, as already mentioned, the millionth component was produced for St. Jude Medical by CarboMedics, Inc. on October 17, 1985.

The St. Jude Medical litigation and settlement opened the door to new thinking and perspectives for the future. We would no longer be a mere supplier of carbon on an OEM basis or a single source supplier of a specialty carbon. We would broaden our scope to include the development of heart valve replacements along with other directly marketed medical devices.

Until now, our medically related activities were a continuation of my extramural excursion into the medical field, characterized by de Hoffmann as my hobby project. The project started out as a department, grew into an entire company division, and evolved to become an autonomous subsidiary of Intermedics, Inc. A bona fide business plan was obviously in order.

We planned to ease ourselves into the market. In the near term, CarboMedics, Inc. would become a supplier of heart valve replacements, first to the People's Republic of China and later to others. Over time, the OEM business would diminish

J. Bokros, *Heart of Carbon*, https://doi.org/10.1007/978-3-031-17933-4_14

and disappear, so starting clinical trials outside of China was the first step. We were led to believe the Chinese would not use a valve without a stamp of foreign approval. This advice came to us from a respected Chinese advisor.

The China valve replacement market was potentially immense. It's reported that 6 in every 1000 Chinese citizens have had rheumatic fever and likely need replacement of their valves damaged by the disease.

In the long term, we'd expand our material's capability to include biomaterials other than carbon, i.e., compatible ceramics and metals (titanium and tantalum). Metals can be useful in the porous form to attach and form a bond with soft tissue and bone, which is useful in orthopedic prostheses. An enterprising engineer already had presented a method to produce porous titanium with a morphology like cancellous bone that was ideal for bonding an orthopedic prosthesis directly to bone. CarboMedics, Inc. had an exclusive license.

In 1979, Michael Jarcho approached CarboMedics, Inc. with a technology that fit right in. Jarcho, who held a PhD in Medicinal Chemistry, developed a methodology to produce hydroxyapatite, a synthetic form of a mineral that is a natural component of bone. In 1981, this led to a venture in dental technology negotiated with Jarcho. The facility was located in San Diego and was chartered as a subsidiary of CarboMedics, Inc.

In 1981, CarboMedics, Inc. formed an orthopedics company. The formation of the orthopedics company was in response to a proposal presented by Nick Cindrich, a seasoned executive with credentials in orthopedics, manufacturing, and marketing at DePuy. A deal was struck. Nick Cindrich became President of the orthopedics company which remained a subsidiary of CarboMedics, Inc.

After the Hemex building was completed, Woerner started on the building for Nick Cindrich, who was responsible for developing orthopedic products that used CarboMedics's proprietary biomaterials.

This portion of the chapter fills in events in the history of carbon beyond the scope of heart valve replacement [1, 2]. Later chapters will focus specifically on advances in carbon heart valve replacements. The CarboMedics group consisted of CarboMedics, Inc. and two others, one dental and the other orthopedic. All three were subsidiaries of Intermedics, Inc.

References

1. Kent JN, Bokros JC. Pyrolytic carbon and carbon coated dental implants. Dent. Clin. North Am. 1980;24(3):465.
2. Bokros JC, Akins RJ, Shim HS, et al. Prostheses made of carbon. Chem. Technol. 1977;7:40–9.

Chapter 15
Two Valves for China

My response to the St. Jude Medical litigation in 1983–1985 was to design two concepts that were similar to the St. Jude Medical patent without actually infringing upon it. In fact, the two concepts were clones.

The PRC-I was a reciprocal pivot, i.e., the pivot was not in a cavity but had the opening and closing pivot stops standing proud, with a rounded protrusion reduced in height, which was positioned between the opening and closing stops. The mechanism resembled a saddle, with the recess between the stops positioned in a notch in the edge of the leaflet. This pivot was similar to that of the St. Jude Medical valve, but outside the St. Jude Medical patent (Fig. 15.1) [1].

The other, the PRC-II, was like the St. Jude Medical valve, but instead of using a spherical recess, the depression was a recessed polygon in which the leaflet tab was captured. The pivot sockets were flared to provide a channel that allowed a backward purge when the leaflet closed, thus clearing the region of stasis (Fig. 15.2) [2].

Both were similar but neither infringed the St. Jude Medical patent. The inside diameters of both valves were identical to the St. Jude Medical.

The PRC-II valve was selected for development. The PRC-II design avoids providing upstream extensions on the orifice "pivot guards," such as those found in the St. Jude Medical. The mandrel process (Fig. 12.4) was used for the manufacture, like that used for the Duromedics valve (Chap. 12). The PRC-I pivot was not purgeable because purging channels were not provided in the design, so it was shelved.

"Call Me Ned"

In the early 1980s, Professor Ned HC Hwang, a chaired engineering professor at the University of Houston, visited CarboMedics, Inc. in Texas.

"Call me Ned," he told me.

J. Bokros, *Heart of Carbon*, https://doi.org/10.1007/978-3-031-17933-4_15

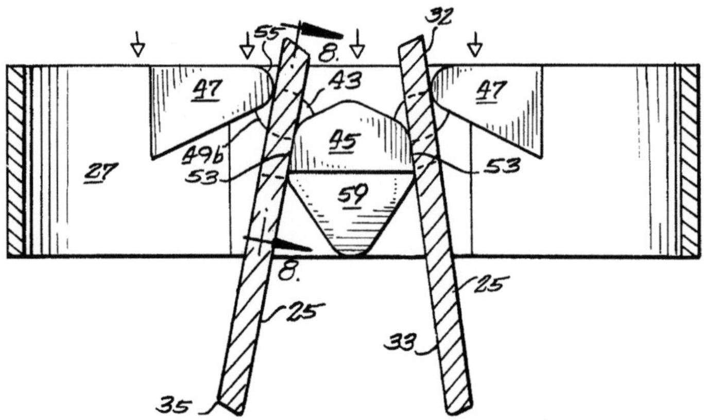

Fig. 15.1 Bokros JC. US Patent No 4,692,165, issued September 8, 1987 [1]

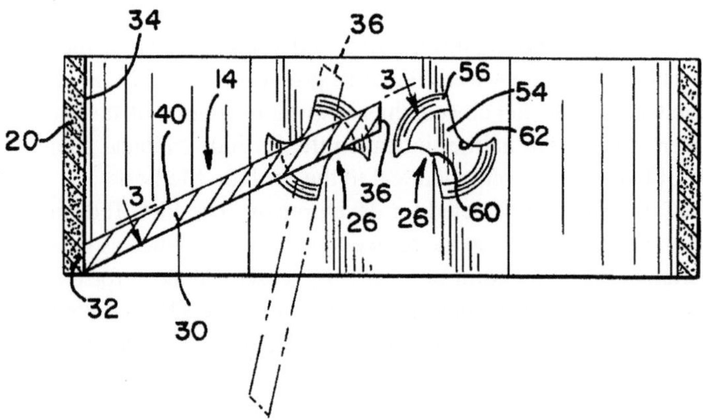

Fig. 15.2 Bokros JC. US Patent No 4,689,046, Heart Valve Prosthesis, issued August 25, 1987 [2]

His specialty was fluid mechanics, focused on characterizing the hemodynamics of blood flow in the cardiovascular system of the human body, including the study of heart valve prostheses. His textbook on fluid mechanics, now in its fifth edition, is still used in more than 50 institutions.

Because Hwang (Fig. 15.3) had connections in China, he became a CarboMedics, Inc. consultant. Hwang's father was special advisor to Chiang Kai-shek's government in China and later in Taiwan was a writer and a poet.

Fig. 15.3 Left to right: Professor Ned Hwang; Dr. Tian Zhipu, Head of CV Surgery Department, West China Medical School (then called the Sichuan Medical School), the first surgeon to perform pediatric open heart surgery in China; Jack Bokros (then President, CarboMedics, Inc., Austin, Texas); and Dr. Ma, Head of the Sichuan Medical School. (Reproduced with the permission of Ned HC Hwang)

Hwang told me he was the Chinese Jack Anderson—I believed he was referring to the columnist, and I don't recall Anderson, but gave Hwang a knowing nod. I learned further that his father was an intellectual, independent, and a bit of a maverick—but he was widely respected.

Our plan was to first gain approval in the Western developed world. After approval, we would negotiate a joint venture with China.

In 1985, Professor Hwang and I traveled to Beijing, Shanghai, Guangzhou, and Hong Kong. Professor Hwang reminded me that I needed to be scrupulous and straightforward to avoid being viewed as a "foreign devil." He and several of his close Chinese associates, including Luo Zheng Xiang, MD, and Prof. Xi Bao Shu, an engineering professor at the Beijing Tsinghua University, provided guidance.

In Guangzhou, I met Lou Zheng Xiang, a distinguished surgeon who also went by CC Luo or just CC. He had trained in Argentina and the United States, was the superintendent of the Guangdong People's Hospital, and was a member of the Chinese Communist Party. As I was later to find out, he was also an entrepreneur (Fig. 15.4).

In Beijing, I met Professor Xi Bao Shu (Fig. 15.5) at the Beijing Tsinghua University. He was a friend of Professor Hwang, with whom he had a close collaboration in developing pulse duplicators for testing prosthetic heart valves. They were ever striving to see whose duplicator duplicated the best.

When Professor Xi asked Hwang for a magnet to be used in his duplicator, Hwang asked me to help carry it to China. It was mounted in a heavy wooden chest (perhaps 35 kilos) with two handles, one for Hwang and one for me.

Fig. 15.4 Left to right: Luo Zheng Xiang MD, Superintendent and Chief of CV Surgery. Guangdong People's Hospital, one of China's heart valve surgery pioneers; Wang I-Shan MD, Vice President and Head of CV Surgery, Shanghai second Medical School; Rolland Siegel; and Jack Bokros. (Reproduced with the permission of Ned HC Hwang)

Fig. 15.5 Left to right: Professor Xi Bao Shu of Tsinghua University and Professor Ned Hwang. (Reproduced with the permission of Ned HC Hwang)

    As a gift for Professor Xi, I brought a bronze belt buckle with a Texas star prominently centered on it. On a subsequent trip to China with Hwang, Professor Xi met us at the train station, proudly displaying the buckle. The star was painted a brilliant red!

    At that time, Professor Xi never spoke English—that's the Chinese way. Later, I found that he really knew English, but Hwang had kept it secret, with a knowing twinkle in his eye. Professor Hwang and his close associates, Dr. CC Luo and Professor Xi, would prove to be crucial assets in implementing the China Plan.

    Until I met Professor Hwang, I did not understand the complexity of Chinese society. I stayed close to Hwang and listened carefully to learn what was or was not proper, including what to say, how to say it, and how to bow. Continuous critiques of my behavior, words, and gestures were freely volunteered.

Rollie Siegel Joins CarboMedics, Inc.

Going forward, anticipating the production and marketing of the PRC-II, CarboMedics, Inc. needed a professional who was experienced in the sales and marketing of heart valve replacements. Rolland (Rollie) Siegel (Fig. 15.6) was identified as the most qualified.

Fig. 15.6 Left to right: Dr. Gu Kaish, Chief Cardiac Surgeon of Shanghai Renji Hospital; Rolland Siegel; Dr. Wang I-Shan, VP and Head of CV Surgery of Shanghai second Medical School; and Dr. Luo Zheng Xiang, Superintendent of Guangdong People's Hospital, Chief of CV Surgery, and a pioneer in heart valve surgery in China. (Reproduced with the permission of Ned HC Hwang)

I met Siegel in 1969 when he was with Cutter Laboratories. Cutter developed valve prostheses for Edward Smeloff, MD; Juro Wada, MD; Nina Braunwald, MD; and Denton Cooley, MD. The Cooley-Cutter valves utilized Pyrolite-coated lenticular discs or hollow bi-conical occluders similar in concept to the hollow carbon balls used in the DeBakey aortic valve.

Siegel had left Cutter in 1975 to join Warren Hancock. He stayed with Hancock Laboratories, focusing on the introduction of the porcine bioprosthetic valve, until the sale of the company in 1980 to the Ethicon Division of Johnson & Johnson for a reported 80 million dollars.

Then Rollie Siegel joined Mickey Ruxin, MD, in a start-up venture, Mitral Medical Inc., serving as an executive vice president. A series of bileaflet mechanical valves was developed by CarboMedics, Inc. for Mitral Medical. These valves opened by swinging *away* from the center line so the primary flow was central, as opposed to St. Jude Medical; CarboMedics, Inc.; and Advancing the Standard valves that swung inward *toward* the center line. The Mitral Medical center-opening bileaflet valve project was terminated during the research phase due to inadequate funding.

I approached Siegel in 1984, after the St. Jude Medical/CarboMedics, Inc. litigation commenced. Siegel was experienced in both tissue and mechanical valve technologies. He liked the CarboMedics, Inc. design due to its solid Pyrolite orifice, the ability to rotate the valve housing within the sewing cuff, its increased leaflet radiopacity, and the absence of any protruding struts or leaflet guards.

Soon after joining CarboMedics, Inc., Siegel resolved a perplexing problem. Although CarboMedics, Inc. personnel were experienced in producing precision-finished carbon parts, they had no experience in designing a user-friendly sewing cuff—the part of the valve that surgeons look to first in judging the merits of a valve design.

Siegel introduced us to a retired lady, Mary Wilson. Mary had been a sewing supervisor at Cutter. She knew exactly what to do. Within several months, we had an elegant and compliant sewing cuff, along with fully trained Vietnamese sewers. Most spoke only Vietnamese. A few were bilingual. Thanks to Siegel, our long-standing problem had been solved.

Siegel liked his independence but agreed to work as a full-time consultant with an annual fee. His first objective was to set up worldwide distribution for the CarboMedics, Inc. valve.

It took less than a year to have European distribution in place. Siegel then sought investigators in the United States and China and a location for a valve assembly facility for a Chinese joint venture. Siegel, lacking experience working with China, collaborated closely with Professor Hwang.

After Professor Hwang, Siegel, and I had completed our first visit to China, and having secured access to Siegel's extensive file of surgeon contacts, we were just one step away from Chinese friends in the United States and Europe.

Our initial thrust was in Europe, to develop clinical data that could be used to obtain the CE mark and then a pre-market approval in the United States. Hwang advised that the Chinese would not accept a valve until it had experience and approval in Europe or the United States.

Siegel was connected with professionals almost everywhere. His list of contacts was remarkably comprehensive: Jim Deegan in the United Kingdom, Olivia Levi in Italy, Mike Barrett in Germany, Andre Hausberg in France, Josep Gatell and Juan Querol in Spain, Roger Lindfors in Sweden, Egon Lovgren and Tom Christensen in Denmark, Kees Feller in Holland, Hans and Sylvia Brady in Austria, Dominique Lamote in Belgium, and Robin Stansen in Australia.

These independent distributors of cardiovascular products had excellent reputations in their respective countries. Together with Siegel's personal contacts, we had access to leading surgeons everywhere.

Siegel and I teamed up. Our modus operandi was for me to lead with the technology presentation. Siegel would then fill in the clinical aspects. He insisted that I be more actively aggressive in marketing the valve.

"You must get comfortable in the front-and-center role," he insisted. This emphasis would be critical to initial acceptance. I took his instructions to heart.

Most important were Siegel's personal files, maintained continually from Cutter to Hancock and onward, documenting his industry contacts. Most everyone we met knew Siegel or had acquaintances in common with him. In the case of young surgeons, Siegel's relationships with the senior surgeon under whom they had trained were helpful in establishing our credentials. Siegel's files reflected his care and personal friendship for all those he met, with notations about their work, hobbies, and families, including wives and children.

I admired the way he engaged people and how he presented the product. He'd sit down, roll out his blue velour cloth on a table with two fingers, look up with a wide smile, and extract the valve with his other two fingers. He would gently give a twist to open and close the valve. Then, with a shake, he'd set it down and sit back with another gratified smile, as if he were presenting a crown jewel (Fig. 15.7).

Siegel directed that the introduction strategy should include invitations to surgeons and distributors to visit our facility so that they'd be exposed to the care and attention to quality that went into every valve. Because of the recently concluded litigation with St. Jude Medical, our staff was initially horrified at the prospect of visitors, but in time they became enthusiastic about presenting their unique workplace and technology. The plant visits ultimately became an important aspect of CarboMedics, Inc.'s sales and marketing program.

First Steps in China

Our first important contact with China came by way of Ned Hwang. Hwang met Zhang Ho (Fig. 15.8), a distinguished Chinese businessman with perfect English, who was based in Houston, Texas. We met casually several times. Zhang was a Chinese buyer who had offices in Houston and worked for "Madam" Wu, who held a high position in the central Bureau of Machinery and Technology in Beijing.

Fig. 15.7 Rolland Siegel presents the CarboMedics, Inc. PRC-II Valve in China. (Used with the permission of Jack Bokros)

Fig. 15.8 Zhang Ho, distinguished Chinese businessman. (Reproduced with the permission of Ned HC Hwang)

When we met, Zhang was purchasing used 707 Boeing aircraft—not one or two at a time, but by the dozen. I recall that upon returning to China, he brought a bottle of Chivas Regal for "Madam" Wu. He said Black Label was better, but she thought it tasted like smoke and chose Chivas instead. Zhang's lunch always included a double shot of Black Label Scotch on the rocks.

Zhang set up our first of many meetings in Shanghai. We met with business figures and then had dinner at a huge circular table (seating 12!) with a rotating center complete with exotic and tasty dishes. After the meeting and a thimble full of mao-tai, Zhang drew up a letter of intent which we all signed. I was suspicious, but Zhang assured me it was not binding: it was just a Chinese tradition.

Zhang arranged another meeting with a Shanghai medical company, which was then making a Björk-Shiley knock-off. I was treated like royalty, first with an extravagant dinner and the next day with a visit to the "shop."

The carbon discs for their valve were produced in a nuclear facility near Lanzhou, a city located in remote Northwest China. All I can say is that the discs were black. They didn't know the crystallite size, density, or preferred orientation. I was not enthusiastic but bowed politely.

Zhang made arrangements to discuss a plan for a joint venture in China. The meetings were held in the Jin Jiang Hotel, an old and prestigious hotel in Shanghai. I came alone to present the business and manufacturing plan put together by Robert Akins, CarboMedics's VPO.

The plan was based on the fact that China and India had a combined population totaling almost three billion people and that six out of a thousand of the population had suffered from rheumatic fever. Assuming that 10%, or even just 1%, would need a valve replacement, the potential need was immense. However, the doctors and implantation institutions required to fill that need were lacking.

The plan was for an initial production facility that could assemble about 20,000 valves per year. Assuming the US export restrictions would not allow us to produce pyrolytic carbon in China, CarboMedics, Inc. would supply rough as-coated components which could then be finished and assembled in China. The valves produced would be for Chinese use and for export to the Far East markets.

The meeting included about 30 Chinese attendees. All appeared of equal rank and each one had things to say; but as far as I could tell, no one seemed to be in charge. One person, who was not introduced, launched into a long, loud monologue in Chinese that struck me as hostile in tone. I surmised that he might have regarded me as a foreign devil (as Hwang predicted might happen). A long dinner was typical; blue crabs were in season, but very messy. Nothing came of the meeting. Had Hwang been there, the result might have been different.

The reception of our plan by the people of Shanghai was not good.

In the fall of 1985, Professor Hwang notified me that he'd been awarded an Honorary Professorship at the second Military School in Shanghai by Dr. Cai Yongzhi, Chairman of Cardiothoracic Surgery at the Changhai Military Hospital.

"I think you should come along," Hwang suggested. "Dr. Cai implanted the first valve in China."

That device was a homemade ball-in-cage valve for a street-sweeping woman. Most important was Dr. Cai's interest in the CarboMedics valve. Hwang told me that Dr. Cai wanted to test the valve in animals. I was looking forward to attending Hwang's award ceremony.

When we landed, Hwang came down with a nagging cough. Cai looked at Hwang's throat and gave him a dose of "Chinese medicine," which cured the cough instantly. After the ceremony, a group of dignitaries was assembled for a photograph (Fig. 15.9a, b).

Fig. 15.9 (**a**) The second Military Medical School in Shanghai presents Professor Hwang with an honorary professorship. Dr. Cai Yongzhi, Chairman of Cardiovascular Surgery at the Changhai Military Hospital, is flanked on his left by Prof. Hwang and on his right by Jack Bokros. (**b**) A cardiovascular surgeon fluent in English provides interpretation. (Reproduced with the permission of Ned HC Hwang)

The military school was interested in conducting animal studies of the CarboMedics, Inc. valve, a study we immediately took interest in. We encouraged the preparation of a plan. Dr. Cai was also interested in the development of a bioprosthesis (Fig. 15.10).

By mid-1986, all of the pre-clinical testing of the CarboMedics, Inc. valve in Europe had been completed. European distribution was ready.

China is a huge country in area and population, with cultures that varied widely by region. In Beijing, we ate "Peking duck." The people were tall and proud. The Cantonese of Guangdong providence are short and loved sweets. This province is the richest in all of China. They did not seem to report to Beijing.

Shanghai is the financial center. The businesspeople resented that all their money was being shipped to Beijing. They liked salty food and blue crabs when in season.

The West and Northwest were mysterious to me. Chengdu is where they had a "carbon" project producing a CarboMedics "knock-off." Hwang had many good friends in Chengdu. During one visit to their carbon facility, he noticed that when their carbon apparatus was switched on, "all the lights in Chengdu dimmed."

I wondered if we'd ever understand the entirety of China. After all, Chinese civilization has existed for more than 4000 years.

Bringing Carbon to Beijing

Our second journey was to Beijing. Hwang and I brought Ernie Lane, an engineer born in Shanghai who, with Dr. Alain Carpentier, had developed the Carpentier-Edwards tissue valves for Edwards to sell.

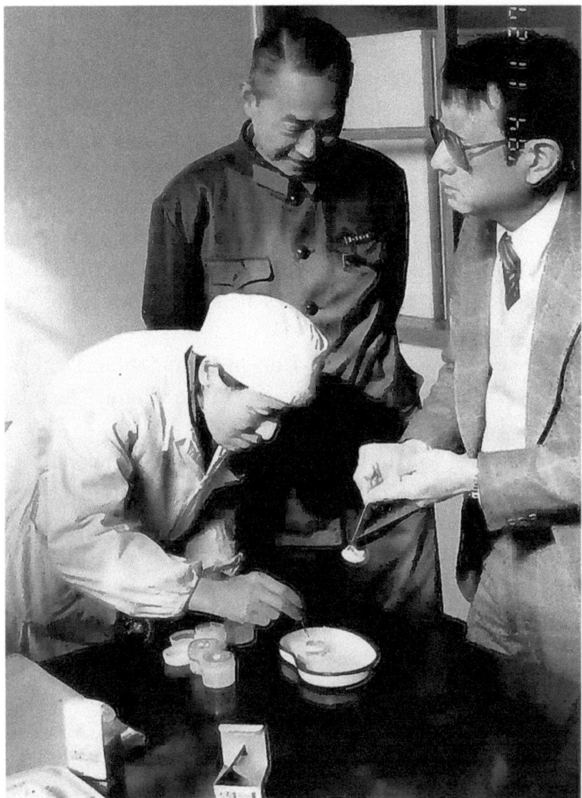

Fig. 15.10 Professor Hwang observes the preparation of a Chinese bioprosthesis. (Reproduced with the permission of Ned HC Hwang)

We first had meetings with about a dozen important doctors and businesspeople associated with Tsinghua University and Fu Wai Hospital in Beijing, then the largest implanting hospital in China. When Hwang was trying to set up the meetings, he told our host in Beijing that we would like to travel to various cities to talk with key doctors. The host said, "No, they will come here." It was a gesture that demonstrated their high regard for Professor Hwang.

One doctor was from Chengdu, where they were trying to duplicate the CarboMedics, Inc. (knock-off) valve. Another, from the Chinese Space Group, attended and said that he was producing the Hall-Kaster valve (knock-off). The doctor from Chengdu said that 30 CarboMedics valves (knock-offs) had been produced and implanted. He said all the patients died. The attendee reporting for the space group was non-committal, but he said the discs often fractured. I came away optimistic that Beijing might be a suitable location for the assembly facility.

After the meeting in Beijing, we boarded a flight to Hong Kong. We took off in an old Russian plane, one with double engines on the wings where US planes had but one engine. We took off to the southeast, but as we were flying, we noticed that the plane was tracking to the northwest. Ernie, always prepared, pulled out a compass and said, "It seems we're going back."

Sure enough, the flight attendant announced in English, "We are going back to Beijing. Buckle up." We were relieved when we landed. Two bicyclists came out to the plane and closed the door. We were told that, all this time, the door had been leaking.

Anyway, the door was properly closed, and we arrived safely in Hong Kong. Ernie Lane (Fig. 15.11), our expert in bio-tissue technology, wanted to tour the slaughter facility that supplied Hong Kong with pork. After the tour, he concluded that they had a big supply, enough valves to last a long time, or as long as the Chinese appetite for pork held out.

After the tour, we met with Dr. CC Luo in Guangzhou. Dr. Luo showed me a graph of the number of implanting institutions in the Chain. The number was small, but the growth rate was exponential!

After 2 years of travel to China, we had enough understanding of the Chinese market and the regulatory aspects to pursue a possible joint venture. We concentrated our search for a location in southern China near Hong Kong.

Consistent with our long-range plans to be active in China, we attended the Chinese International Cardiovascular Surgery Congress being held Oct. 27, 1986, at the Fragrant Hill Hotel, a resort about 40 miles from Beijing. As this was our first

Fig. 15.11 Left to right: Dr. CC Luo Zheng Xiang; Ernie Lane; Jack Bokros; and Professor Hwang. (Reproduced with the permission of Ned HC Hwang)

opportunity to exhibit the valve intended to become a "valve-of-choice" in China, this meeting generated a maximum logistical effort on our part. We carried exhibit materials from Austin along with plenty of support staff.

On arrival at the hotel, Siegel and I were surprised to discover that attendees included leading surgeons from the United States, such as Dr. John W. Kirklin from the University of Alabama and Dr. John Collins from the Harvard-affiliated Brigham and Women's Hospital. Spontaneous contacts with these leaders, literally over the breakfast table, eventually led to the adoption of our valves at their centers.

During exhibit hours, the aisle in front of our booth was jammed with curious Chinese surgeons, all of them wanting to see the latest in US valve technology. In his usual meticulous way, Siegel gathered names and addresses. We noted that many of the Chinese surgeons had extravagant titles, perhaps to compensate for the modest circumstances of their hospitals.

For this occasion, the Congress invited Dr. Robert Jarvik and his fiancée, Marilyn vos Savant, a widely read syndicated newspaper columnist (*Ask Marilyn*), to China so Dr. Jarvik could present a plenary overview of artificial hearts and support devices. He also issued some favorable remarks about the newly developed CarboMedics, Inc. heart valve replacement.

Dr. Jarvik and Marilyn were great fun to be around. He fondly took pride in his popular and well-known fiancée, constantly circling her to take rapid fire photos. His childlike joy in having fun showed itself at dinner. He was seated at a nearby table, lowered his head into a napkin in anticipation of a sneeze, but instead reared up with a pair of ivory chopsticks wedged in his upper jaw, portraying an aggressive walrus. The young Chinese waitress was in shock and amazement at the antics of this famous American surgeon.

Dr. Jarvik was fascinated with CarboMedics Pyrolite Carbon and asked us to manufacture wedding bands of Pyrolite Carbon. We supplied the carbon blanks, and Dr. Jarvik had simple bands of a very thin gold delicately inlaid on their rims, highlighting the shiny black bands. We also made a wedding gift of Pyrolite Carbon napkin rings, presented to the couple in a handmade teak chest.

Robert and Marilyn were later married in an extravagant ceremony at the New York Plaza Hotel at the end of Central Park (Fig. 15.12). Noted science fiction author Isaac Asimov gave the bride away.

Anticipating the introduction of the CarboMedics, Inc. valve in the United States, Dr. Jarvik arranged a formal dinner in Austin. Jarvik, the President of Symbion, Inc. (an artificial heart company), sat at the end of the dinner table. He clinked his wine glass to get everyone's attention and started to regale us about the capabilities and advantages of having him and Symbion represent CarboMedics in the exclusive marketing of the CarboMedics valve.

We took about a week or so to dispassionately analyze Symbion's capabilities. We decided that we had to decline his offer, even though we would have enjoyed working with such a dedicated individual. I could tell that he was disappointed by our response, but his friendship continued unabated.

Fig. 15.12 Jack and Roberta Bokros at the wedding of Robert Jarvik and Marilyn vos Savant at New York Plaza Hotel. (Reproduced with the permission of Robert Jarvik)

First Human Implant of the CarboMedics, Inc. Replacement Valve

We were eager to complete our first human implant prior to the end of that calendar year. Rollie Siegel identified two prominent surgeons, both of whom agreed to be initial implanters: Professor J Y Neveux in Paris and Professor Roland Hetzer in Berlin. Siegel, Chuck Griffin, and Gwyn Baumgarten (a former operating room supervisor for Denton Cooley) flew to Berlin with a small inventory of valves. Upon arrival, they were informed by Professor Hetzer that a suitable patient was not on the schedule.

Stymied, they went to Paris, where a suitable patient was scheduled for aortic replacement. Neveux selected a replacement candidate. She was a Paris resident who would be available for long-term follow-up. The replacement surgery, performed on December 18, 1986, was uneventful. Our entry into the direct competitive valve arena was under way.

Shortly after the first implant in 1986, Siegel, Hwang, Neveux, and I traveled to Japan, Hong Kong, and Guangzhou in Guangdong province to celebrate the first CarboMedics valve implanted in Paris by Dr. Neveux.

The first stop was Tokyo, a courtesy call on Dr. Hitoshi Koyanagi, the leading implanter of the St. Jude Medical valve in Japan. Our discussions, with Neveux taking the lead, received a smile of acknowledgement, pleasantries, and a respectful departure. We were not really expecting anything more, as Koyanagi's relationship with St. Jude Medical was cozy (Fig. 15.13).

Fig. 15.13 Left to right: Professor Neveux; Dr. Koyanagi with a colleague; Jack Bokros; and Rolland Siegel. (Used with the permission of Jack Bokros)

In Hong Kong, we were greeted by David L C Cheung, MD, together with two others. Professor Samuel Hulbert and his wife, Joy, just happened to be on a tour of Southeast Asia and attended David Cheung's meeting. Sam has been identified as a "father of biomaterials" and a researcher of Pyrolite Carbon.

David's meeting was a gathering of local Hong Kong dignitaries, including Mok Hing Yiu, MD (who had fallen ill and could not be present), and several regional surgeons. I was privileged to introduce Dr. Neveux (Fig. 15.14), the first implanter of the CarboMedics valve that had been designed for China.

After Dr. Neveux's presentation, David Cheung took our group to dine at the exclusive Jockey Club and explained, "My father sells biscuits. If everyone in China had a biscuit at breakfast, that's a lot of biscuits"—hence the Cheung Dynasty. He volunteered, with a laugh, "All my brothers are important businessmen. I'm just a lowly cardiac surgeon."

We then traveled to Guangzhou to secure a meeting between Drs. Neveux and CC Luo about providing the CarboMedics valve to China and Asia via a joint venture.

Hwang elected to travel in the main cabin of the ferry. His journey was, in Chinese, through the "mouth of the snake" with the locals, many with livestock. Siegel and Dr. Neveux traveled by train and were transported to the new Garden Hotel that replaced the ancient Hang Fang Hotel, where I had once stayed courtesy of the Shanghai medical company.

Dr. CC Luo made the reservations. Even though Dr. Luo was a principal in the People's Congress, he had a bent toward capitalism. His affinity for capitalism was evidenced on a motor trip from Guangzhou to Zhuhai. Passing through a pretty wooded area of commercial fishponds, he declared with an outward sweeping

Fig. 15.14 Jack Bokros and Dr. J Y Neveux first implanter of the CarboMedics valve. (Used with the permission of Jack Bokros)

gesture, "We could build villas here." I never forgot that as we passed the South China Sea to survey a possible assembly site in Zhuhai.

Earlier in 1986, Dr. Luo visited CarboMedics in Austin. Picked up and transported by the Intermedic's Learjet, he stayed in our guest house situated in the midst of towering pecan trees. The first morning, when I walked up to the guest house, he was sitting on the porch. As I approached, he gave me another of his sweeping gestures, saying "a great place." His wife, a cardiologist, resided off and on in Santa Barbara, California, an even prettier place. The Luos were well-connected Chinese dignitaries.

When we arrived in Guangzhou, Luo put me in the presidential suite on the penthouse floor of the Garden Hotel. All of the bathrooms had ceramics, with images of exotic fish lining the toilet and bidet. Luo had put out the word that there would be an important symposium at the Garden Hotel. In response, the meeting room was filled with regional surgeons and cardiologists.

The night before, we met at dinner, where Dr. Luo set the agenda. Dr. Neveux would describe his experience with the "Chinese" valve (as he now called it). Then I would present the carbon technology.

In the morning, Luo introduced Dr. Neveux in Chinese. Neveux made his presentation in English, with Luo occasionally breaking in to comment in Chinese. It took about half an hour.

I went to the podium and said good morning. Turning to the screen, I pressed the control on the slide projector. It hummed and shook and finally began to expel slides up and around.

"Good lord," I think I said.

Hwang jumped up, telling me in English, "Sit down!"

In Chinese, he launched into what I imagined, by the claps and burst of laughter, was a complete presentation without slides that pleased everyone present—but no one more than me. It couldn't have turned out better if I'd rigged it myself.

Luo joked afterward, "Your presentation was well received!"

Thanks to Professor Hwang.

On a subsequent visit to Hong Kong in 1987, as my wife Roberta and I entered the lobby of the Sheraton Hotel, we were engaged by a freelancer looking to help outsiders join with China in joint ventures. His name was CC Tam (not to be confused with Dr. Zheng Xiang Luo, known to me as CC of Guangzhou). CC Tam was Chinese, but fluent in English. His associate was Andrew Tsui, a radiologist who also was fluent in English.

Both were broadly knowledgeable of business development in Hong Kong. When I was introduced to Andrew, he presented me with a gift, a grain of rice with the *Lord's Prayer* engraved on it. It was sealed in a small container with a magnifier on top so the words were clearly visible. This small gift made an impact, a gesture long remembered. Zhang Ho, whom we met in Texas, never expected payment. CC Tam and Andrew were our first paid contacts in China, hired as consultants.

References

1. Bokros JC. US Patent No 4,692,165, Heart Valve, issued September 8, 1987.
2. Bokros JC. US Patent No 4,689,046, Heart Valve Prosthesis, issued August 25, 1987.

Chapter 16
Negotiating Chinese Joint Venture

In 1987, we settled on an area in or near Hong Kong for the assembly facility site. Hong Kong was booming with commerce. Two sites were identified.

One was an unfinished building designed to be an upscale restaurant located on the shores of Hong Kong Bay. It was two stories tall with a huge dining area on top that could house the main assembly and inspection area. The vibratory polishing equipment could be on the ground floor. The quote for the cost of a lease seemed very low, so I took it with a grain of salt.

The other site was the more attractive option. It was in an industrial area, a concrete and steel block building, perhaps 20 stories high, with a large rectangular area on top measuring about 100×120 feet, all open but ready for a finish out. Akins was excited about it. There were a large parking lot, an ornate stainless steel and marble ground floor lobby, and an elevator that went straight up to the top. It was quite fancy. All the walls were windows.

Our negotiating team included Akins and me from Austin, Siegel from Portland, CC Tam and Andrew Tsui (our consultants from Hong Kong), Dr. David Cheung, Dr. Luo, Professor Xi from Beijing, Professor Ned Hwang, and Zhang Ho.

On the Chinese side, there were a half dozen local government officials from Hong Kong, the Shenzhen and Guangzhou free zones. This included Hwang's cousin, who had been the Chairman of the Guangzhou Economic Development Zone in Guangdong Province.

It was clear that Guangdong Province was endowed with a wide variety of skilled workers in electronics, textiles, and manufacturing. It was a main port entrance into China, and the Province "did not take orders from Beijing."

We felt up to speed with the regulatory rules and had a fairly good idea about leasing costs. Technical personnel trained in electronic assembly were available; labor cost rules were complicated, with many regulations that were tedious but manageable. Akins had a plan for the venture facility and was satisfied with the utilities.

J. Bokros, *Heart of Carbon*, https://doi.org/10.1007/978-3-031-17933-4_16

Fig. 16.1 Discussing the joint venture proposal. CC Tam; Dr. Luo Zeng Xiang; and Jack Bokros. (Used with the permission of Jack Bokros)

We were ready to talk and discussed a joint venture proposal (Fig. 16.1). Akins detailed a kit that contained all the parts for assembling a valve. The carbon parts would be unfinished, but manufactured to CarboMedics, Inc.'s specifications, ready to be finished. The kit would include metallic assembly rings and a main titanium outer band, together with the cloth for the sewing cuff complete with sutures.

Our estimate of the minimum transfer cost was in the $400 to $500 US range.

At the same time, back in Texas, there were rumbles about a sale of Intermedics, Inc. (which would include CarboMedics). We had no control over that and decided to stay on task and not worry until such a sale had actually occurred.

We met with all parties on the Chinese side and were ready to get serious about a joint venture. Putting all the other logistics aside, the important factor was the cost of a kit that included everything required to make a valve.

To minimize the kind of negotiating time the Chinese were famous for, Akins set our lowest cost for the kit and knew the minimum cost per set below which we could not go. It was $479 US, but this was not to be disclosed until the next morning. If the Chinese price requirement was below our minimum, "we were out of there." The Chinese side would determine the maximum price the joint venture could pay to be viable in China.

Morning came. I had a cost of $479 US fixed in my mind yet to be disclosed. The Chinese were ready to disclose their maximum price. It was a tense moment when push would come to shove.

Their announced price, believe it or not, was $479.

I was dumbfounded. $500 would have been believable, but $479 on the nose? Someone on our team was playing both sides against the middle. An exact price match was outrageous.

Even though the negotiation had just begun, I ended the meeting there and then, saying we'd have to think about it. We hurried back to Texas to see about the rumored sale.

As for us getting back to China? It never happened.

Chapter 17
European Company Buys Intermedics Inc.

In 1988, the announcement of the sale of Intermedics, Inc. to a European conglomerate for a total of about $800 million came abruptly, out of the blue. For CarboMedics, Inc., it was utterly disruptive. The nature of the negotiation between Intermedics, Inc. and the European company was held close to the chest and, looking back, was similar to the negotiation to sell General Atomic's Medical Products Division to Intermedics, Inc. Once again, I was kept in the dark until the deal was done.

The CarboMedics, Inc. group had taken sales from $4 million in 1979 to an anticipated $90 million through 1988, including the sales from its two start-ups, the dental company, and the orthopedics company.

It was rumored that about 50% of the total sale price was for the CarboMedics group that included the dental company and the orthopedic company. The other 50% was for the Intermedics, Inc. pacer company and all other assets.

Intermedics, Inc. price for the Medical Products Division was six million dollars, plus some Intermedics, Inc. common stock. The management team (Bokros, Akins, Haubold, Emken, and Sommerfeld) signed employment agreements that included 4% of gross sales as an incentive. Each had received phantom shares of Intermedics, Inc. common stock. The term of the agreement was 10 years. In 1988, when the employment agreements expired, so did the patents that protected the carbon technology.

The abrupt breakup of Intermedics, Inc. dealt a shock to its individual subsidiary companies.

The autonomous subsidiary, CarboMedics, Inc., and its two associate entities stayed intact. However, the five executives, including me, who had been contract employees with 10-year employment contracts due to expire at the end of 1988, had already been considering our options.

J. Bokros, *Heart of Carbon*, https://doi.org/10.1007/978-3-031-17933-4_17

After reviewing a proposal for a 10-year contract set forth by Intermedics's broker for the sale to the Europeans, the management team decided that we'd be better served by letting our current contracts lapse so we could leave to pursue other endeavors. In such a case, the top management that remained agreed to close ranks and come up to speed, as the business was ongoing without serious disruption.

After the sale, GR Chambers, the outgoing President and CEO of Intermedics, Inc., approached me with the idea of making an initial investment of $5 million for any new venture, with more available, depending on our performance. I thanked him for the offer, telling him I'd get back to him.

The orthopedics company, an affiliate of CarboMedics, Inc., was established in the early 1980s with Nick Cindrich as its President and CEO. Cindrich and his team stayed through the sale, but became uneasy, so in 1992 they elected to leave and create a new company, Encore, Inc. After 4 years of operation, Encore made headlines in 2006 when a private equity entity, Blackstone Capital Partners, bought Encore for $870 million, one of the highest prices ever paid for an Austin company.

After the departure of Cindrich and his associates, the Intermedics orthopedics company continued to be run by the European company's existing orthopedic division. In the early 1990s, the orthopedics sales amounted to about $180 million.

The dental company, which had 125 employees and was generating about $30 million in US and foreign sales in 1989, remained as a European company.

There were a number of unfinished projects that were included in CarboMedics, Inc. whose fate was uncertain. One of these involved a wide-ranging use of percutaneous devices that included through-the-skin ports for electrical leads and/or ports for fluids, as in dialysis procedures. Another included dental implants used to support individual teeth or bridges, as well as carbon devices that attached directly through the skin into the skeletal bone to directly support an artificial limb. Details of all these have been published [1–3]. Obviously, the envisioned joint venture with the Chinese to assemble and sell a bileaflet Pyrolite Carbon valve in China had become just such an orphaned project.

At this point in time, Robert Akins (Fig. 17.1), Vice President of Operations of CarboMedics, decided to retire and return to California. Akins had struggled with the Texas climate for 10 years. He was ready to retire and enjoy life. I was disappointed because Akins had been the key contributor since General Atomic days.

I wasn't surprised that Akins, who was a planner, had nurtured Jonathan Stupka (Fig. 17.2), who was happy to "step right into Akins's shoes."

During the lull after the sale of Intermedics, Inc., my colleagues and I wrote two chapters [4, 5] for *Replacement Cardiac Valves* edited by Endre Bodnar and Robert Frater, published by Pergamon Press [6]. The book, in its Chap. 13 by Benson B Roe, contains a very interesting and complete historical record of the development of mechanical cardiac prostheses.

Fig. 17.1 Robert J Akins, Vice President of Operations of CarboMedics, Inc.. (Used with the permission of Jack Bokros)

Fig. 17.2 Jonathan Stupka stepping into the shoes of Robert Akins. (Used with the permission of Jack Bokros)

References

1. Klawitter JJ, Patton J, More R, Peter N, Podnos E, Ross M. In vitro comparison of wear characteristics of pyrocarbon and metal on bone: shoulder hemiarthroplasty. Shoulder Elbow. 2018;12:11–2.
2. Kent JN, Bokros JC. Pyrolytic carbon and carbon coated dental implants. Dent Clin N Am. 1980;24(3):465.
3. Bokros JC, Akins RJ, Shim HS, et al. Prostheses made of carbon. Chem Technol. 1977;7:40–9.
4. Bokros JC, Haubold AD, Akins RJ, et al. Trends in prosthetic heart valve design in. In: Bodnar E, Frater RWM, editors. *Replacement cardiac valves*. New York: Pergamon Press, Inc.; 1991. p. 333–55.
5. Bokros JC, Haubold AD, Akins RJ, et al. The durability of mechanical heart valve replacements: Past experiences and current trends in. In: Bodnar E, Frater RWM, editors. *Replacement cardiac valves*. New York: Pergamon Press, Inc.; 1991. p. 21–48.
6. Book entitled Replacement Cardiac Valves, Book eds by Endre Bodnar and Robert Frater. Pergamon Press USA, copyright 1991

Chapter 18
Medtronic Inc. Project

In 1988, shortly after the sale of Intermedics, Inc., the president of Medtronic, Inc.'s heart valve company expressed the company's interest in starting a new carbon manufacturing company with the capability of developing a Pyrolite Carbon bileaflet valve that they could sell. At first, I rejected the proposal. All of the management team wanted to be independent. We were all leery of working for large corporations after our joint experience of being unexpectedly sold.

Medtronic, Inc. followed up by asking me to come to Minneapolis to discuss the proposal. I went without an attorney and met with the Medtronic, Inc. executives, including several attorneys. As I was without counsel, a Medtronic, Inc. Executive Vice President dismissed their attorneys. Perhaps the Executive Vice President thought I was there to make a deal, not just to listen.

At the beginning of our discussion, I related that I already had financial support and had no time to waste. The Executive Vice President offered a "$200,000 for a 90-day standstill" arrangement. After some discussion, he increased the offer to $600,000, to be applied to any deal that might be forthcoming.

The Executive Vice President then opened up, saying that Medtronic, Inc. was prepared to fund everything. They wanted a carbon facility that could produce 40,000 bileaflet valves per year and a valve to be designed and developed.

That statement turned me around. I agreed to negotiate an agreement.

Upfront, it was made clear to Medtronic, Inc. that I would not consider permanent employment, but if a successful project was completed with them, I would be free to depart to compete in any field, including heart valve replacements using the existing carbon technology as well as improvements to that technology achieved during the course of the project.

1. The Project would be carried out by a company named Carbon Implants, Inc. including its employees.
2. Carbon Implants, Inc. would own the existing carbon technology and any improvements to the processes or heart valve design patented during the project.

J. Bokros, *Heart of Carbon*, https://doi.org/10.1007/978-3-031-17933-4_18

3. Any Bokros Entity (a term defined by Medtronic, Inc. to represent the group that performed the Project and survived the sale of Carbon Implants, Inc. to Medtronic, Inc.) would be able to use the same carbon technology, and any improvement to the technology made during the project with Medtronic, Inc., to produce any product the Bokros Entity desired, including heart valve replacements.
4. The Project would end when the valve design was frozen, termed the "Design Freeze." Medtronic, Inc. would carry out clinical trials.
5. Employees (22 persons) of Carbon Implants, Inc. would sign new agreements, allowing them to participate in any Bokros Entity.
6. Bokros Entity employees would be available for a time to help Medtronic, Inc.'s new facility produce the new heart valve that would be marketed by Medtronic, Inc.'s Heart Valve Division.
7. I proposed to hire a co-worker at General Atomic, a PhD operationally familiar with the carbon technology, to train personnel to run any project that Medtronic, Inc. might continue.

Medtronic, Inc. expressed concerns but finally agreed, provided that:

1. The venture wouldn't be sold before a fixed number of years.
2. I would be in control for the fixed number of years. The entity that would be called the "Bokros Entity" included every venture we decided to pursue.
3. No Bokros Entity would be sold to another company that had been in the heart valve business for a specified time before the sale.
4. In the event of a sale after the fixed number of years, Medtronic, Inc. would have a preferred advantage in purchasing the venture.

When the deal was done, we leased a building (Figs. 18.1a–c). The facility was designed to produce 40,000 bileaflet carbon valves per year.

Then a diversion came. We received a demand letter from CarboMedics, Inc. dictating just what we could or could not do with respect to CarboMedics, Inc. proprietary property. Bob Akins and I, at the suggestion of GR Chambers, visited an attorney located near Intermedics, Inc. in Angleton, Texas, a contingency attorney.

The contingency attorney prepared a suit asking for a declaratory judgment action that would force CarboMedics, Inc. to state exactly what we could and could not do. The suit with CarboMedics, Inc. was settled in 1991, with no monetary award but certain restrictions on new hires—kudos to the contingency attorney.

Initially, we planned to develop a valve concept that Ned HC Hwang had presented to me as a design to be pursued by Carbon Implants, Inc. A dispute over the ownership of the concept precipitated a lawsuit initiated by the University.

When the University of Houston suit settled, the University paid our side $400,000. Prof. Hwang's concept was abandoned. A new design was pursued.

On February 24, 1991, I turned 60 years old, but was not dwelling on that milestone. I had been working in my office when the intercom announced, "Dr. Bokros, come to the coating room" where the 40,000 valves per year facility was under

a   b

c

Fig. 18.1 (**a**) Visiting dignitaries to the new Carbon Implants, Inc. facility. Left to right: Luo Zheng MD, Chief of CV Surgery of Guangdong Peoples Hospital, and Professor Ned Hwang, University of Houston (Reproduced with the permission of Ned HC Hwang). (**b**) Announcement of the formation of Carbon Implants, Inc. (**c**) Management team, relaxed and toasting the formation of Carbon Implants, Inc.. (Used with the permission of Jack Bokros)

construction. I opened the door to be met with a flash of light and a loud "Happy Birthday." Surprised I was! The room was full of about 100 people from the past.

The highlight of the event was Ned Hwang's unveiling of a bust he had commissioned from renowned artist Willy Wang (Fig. 18.2).

I was very pleased of course, honored by Hwang's remembrance, but humbled by Willy Wang (prominent for sculptures of many Chinese dignitaries, as well as the huge sculpture honoring China's workers at the front corner of Tiananmen Square). In Fig. 18.3, Hwang is on the left and Willy Wang on the right.

GR Chambers, who had been my boss at Intermedics, Inc./CarboMedics, also attended the celebration with his wife (Fig. 18.4).

There were many mementos. I'll show only one. It was a piece of glass, 20″ by 33″, framed with a light in the base to reveal a long sweeping scratch. Michael Emken (Fig. 18.5) had somehow acquired my 1968 office windowpane, the one

Fig. 18.2 Professor Ned Hwang presents bust to Jack Bokros. (Reproduced with the permission of Ned HC Hwang)

Fig. 18.3 60th birthday party Professor Ned Hwang and Sculptor Willy Wang. (Reproduced with the permission of Ned HC Hwang)

Fig. 18.4 60th birthday party Jack Bokros; GR Chambers and his wife Rita. (Used with the permission of Jack Bokros)

Fig. 18.5 The brass plate attached to the framed glass presented by Michael Emken. (Used with the permission of Jack Bokros)

bearing that distinctive scratch. The scratch was made when the first Pyrolite Carbon was produced and handed to me. I had promptly walked up to that window to test the sample's hardness by trying to scratch the glass.

A good time was had by all.

The design of the new Medtronic, Inc. valve incorporated a long straight orifice, substantially longer than typical of the CarboMedics, Inc. and PRC-I valves. The leaflets and orifice were designed to allow the leaflets to open to 90°, parallel to the flow (Fig. 18.6) [1]:

United States Patent

Stupka et al.

Patent Number: **5,192,309**

Date of Patent: **Mar. 9, 1993**

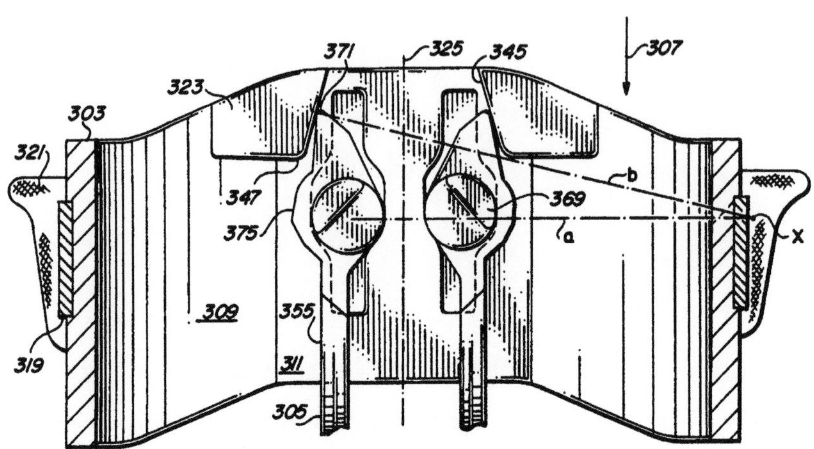

Fig. 18.6   US Patent No 5,192,309, Prosthetic Heart Valve, issued March 9, 1993

1. In the open position, the tips of the leaflets rest at the bottom of the sloping surfaces on the abutments. The sloping surfaces are "closing ramps."
2. The closure is actuated by "closing ramps."
3. On flow reversal, the leaflets move upward, sliding along the ramps. A camming action forces the leaflets toward the closed position.
4. Reaching the closed position, the raised portion of the pivot (#369 on the patent figures) rotates so the raised ledge is perpendicular to the flow.
5. The movement of the pivot with the flow provides a wiping purge to minimize stasis.

Interestingly, in this same time frame and as another diversion from Carbon Implants, Inc., a newly formed company, Advancing the Standard Medical Inc., contacted me looking for a carbon valve replacement design that they could commercialize.

They were advised that we had already made a commitment to Medtronic, Inc./Carbon Implants, Inc. and could not help, but suggested that CarboMedics had a valve patent on the shelf that was outside the St. Jude Medical patent.

The patent was the PRC-I depicted in Fig. 15.1. Reportedly, Advancing the Standard Medical Inc. paid $41 million to CarboMedics, Inc. for the patent and the mandrel technology (Fig. 12.4) needed to produce the valve. There may have been a royalty. Subsequently, Advancing the Standard Medical Inc. obtained an improvement patent (Fig. 18.7, Hanson, et al.).

Eventually, Medtronic, Inc. acquired Advancing the Standard Medical Inc. and continues to sell the Advancing the Standard valve prosthesis.

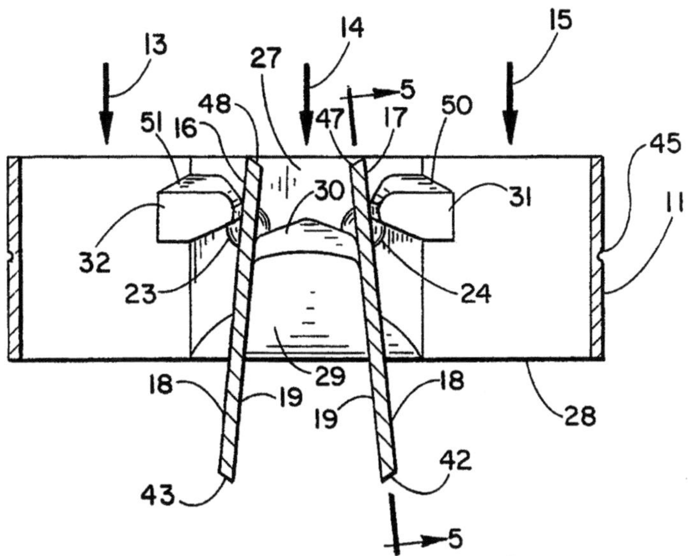

Fig. 18.7 US Patent No 5,354,330, issued October 11, 1994, is subordinate to PRC-I patent

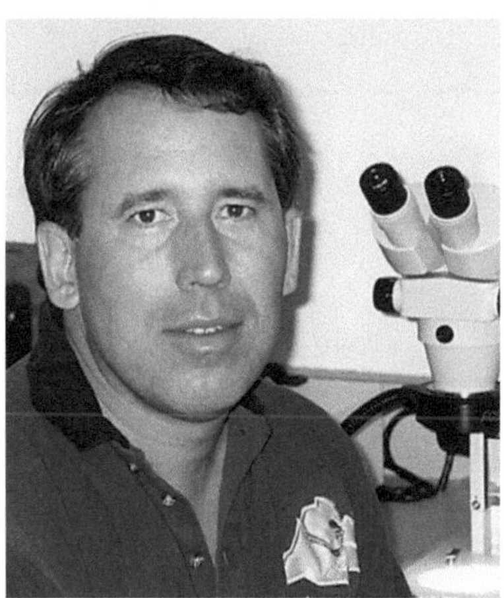

Fig. 18.8 John Ely, VP of Regulatory Affairs. (Used with the permission of Jack Bokros)

After the diversion, John Ely (Fig. 18.8), who had been employed by Hemex, Inc. (Chap. 12), applied for the Carbon Implants, Inc. lead regulatory position. He had taken Hemex, founded in the early 1980s, through the entire regulatory process, received pre-market approval, and proceeded onward to commercialization.

Edwards Laboratories purchased Hemex in 1986 and relocated it to California, along with John, who signed an employment agreement and moved. Because of his wife Patty's allergies, John applied for employment. Texas is a right-to-work state, so I hired him. It took 3 months of negotiations for the matter to be settled … while John played golf. The settlement had temporary restrictions, but John was soon named Carbon Implants, Inc.'s VP of Regulatory Affairs.

A Key Technical Breakthrough

About 2 years into the project, David Wilde, a carbon processing technologist, came to me with an idea. He suggested a precise methodology that used a materials balance approach to create a steady state during the carbon deposition process that would allow control of the size of the fluidized bed. I told him we had already tried his approach, but suggested he give it a try.

Within a month, Wilde and Jim Accuntius showed me a plot of bed surface area versus time (Fig. 18.9). Their results indicated that the control was working so well that it would now be possible to coat to precise dimensions and "coat-to-size." [2] Wilde remarked politely, "Whatever you did before, you didn't do it right."

Pleased, I asked him, "How many shares of the company do you have?"

He told me, and I said, "Now you have double that."

Then John Ely picked up the ball and tediously carried out parametric studies, mapping out the relationship between the deposition condition and the strength of

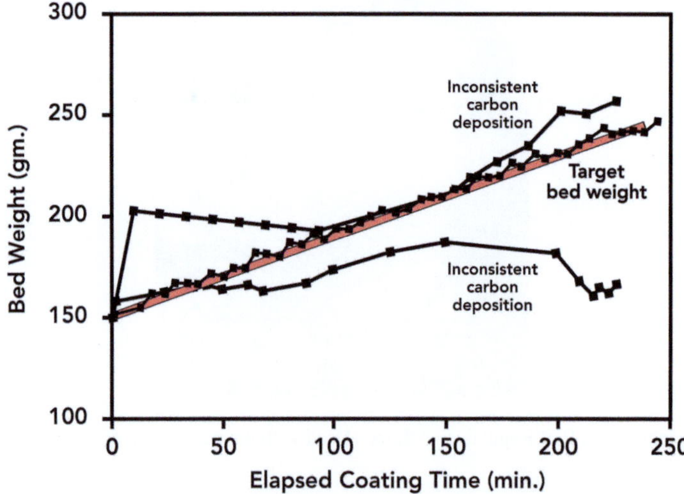

Fig. 18.9 Bed surface area versus time for carbon-coated components. Controllable parameters mean that results are now reproducible. (Reprinted from the *Journal of Heart Valve Disease*. Pure pyrolytic carbon: preparation and properties of a new material, On-X® carbon for mechanical heart valve prostheses, JL Ely, et al., 1998 7;6 with the permission of ICR Publishers Ltd)

isotropic carbon to determine the conditions needed to produce the strength and toughness desired. He filed a patent [3] to make those conditions proprietary and published his findings [4]. By agreement, this technology was owned by Medtronic, Inc., but with a "right to use" assigned to any Bokros Entity.

We showed the control to Medtronic, Inc., suggesting the new carbon would improve the new 1205 valve, to be named Parallel. Word came back from on high, "The Board thought it was too risky." The Medtronic, Inc. Executive Committee recommended staying the course with the proven Pyrolite Carbon.

Medtronic, Inc. insisted that once the conditions for "Design Freeze" were achieved (when all the hemodynamic, durability, and animal data had been completed by hand-picked experts noted for their technical prowess and then thoroughly reviewed), the contractual milestone was fully satisfied. A meeting of all the investigators, consultants, and the Medtronic, Inc. Executive Committee gathered in a huge formal meeting room, recorded by video and audio equipment, to declare "Design Freeze" with a formal toast. Engraved mementos were prepared for commemorating the occasion.

When the design was frozen, then and only then could Medtronic, Inc. take over and exclusively carry out the clinical trials. At this point in time, the Carbon Implants, Inc. project was complete, and all the requirements had been satisfied, so the sale to Medtronic, Inc. was set to close. The deal included a manufacturing facility capable of producing 40,000 Pyrolite Carbon bileaflet valves per year for Carbon Implants, Inc.

In addition to a substantial cash payment, Medtronic, Inc. agreed to the payment of a royalty, together with a series of bonus payments due when certain objectives were met. A royalty stream for the benefit of participants in Carbon Implants, Inc. was anticipated.

The cash payment was made, with 10% holdback for 2 years, and this was eventually paid.

Then came a rebound. Shortly after Design Freeze was declared, the president of Medtronic, Inc. paid me a visit. He arrived very stiff and stern. We sat down at my long, old, marble-topped table (the one that John Ely had cracked when he sat on a corner of it). The President waved a white sheet of paper in his left hand (the image still remains etched in my aging mind).

Shaking the paper, he demanded, "What have I bought? What have I bought?" He slid the paper across the table toward me. I saw that it was the Carbon Implants, Inc. organizational chart with 22 names circled on it. He glared at me as I recoiled, thinking how to respond. Finally, I said slowly and softly, "That was the deal."

He departed without another word.

Shortly after the Medtronic, Inc. clinicals started in Europe in February 1995, clotting of the Parallel Valve occurred, and Medtronic, Inc. provided a report to us on three thrombosed valves. When we reviewed the Medtronic, Inc. protocol, we found that the Medtronic, Inc. clinical trial was marked confidential. That meant the data forthcoming were hidden from disclosure. No further comment is possible.

Reports of thrombosed valves continued. Looking forward as a consequence, the credibility of Medical Carbon Research Institute was in question.

The Accuntius/Wilde Patent Makes a New Pure Carbon Valve

We elected to pursue the Accuntius/Wilde patent [2], described below, that made it possible to design a valve replacement based on hemodynamic theory. Such a design could deliver performance approaching that of the natural valve, replete with low turbulence and little to no blood damage.

All carbon heart valve replacements between 1968 and 1992 used Pyrolite components, which were produced in a fluidized bed (Fig. 5.2 in Chap. 5). Attempts to establish a steady state by maintaining a material balance failed to achieve good control of the bed surface area. The process was not precise enough. However, the new control system provided the needed control (Fig. 18.10).

Gaining proper control represented a paradigm shift in pyrolytic carbon processing in fluidized beds. Precision control made it possible to eliminate the silicon carbide from Pyrolite Carbon. This in turn restored the precision necessary to use the superior carbon to produce a pure carbon valve replacement that was to deliver features never achievable with the older Pyrolite process.

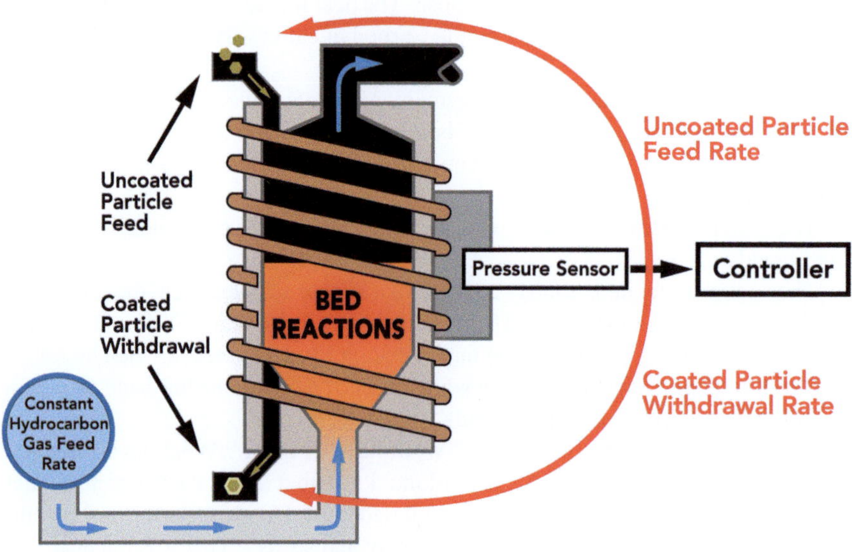

Fig. 18.10   The bed control apparatus. (Reprinted from the *Journal of Heart Valve Disease*. Pure pyrolytic carbon: preparation and properties of a new material, On-X® carbon for mechanical heart valve prostheses, JL Ely, et al., 1998 7;6 with the permission of ICR Publishers Ltd)

References

1. Stupka JC, Bokros JC, Emken MR, Haubold AD, Peters TS. US Patent No 5,192,309, Prosthetic Heart Valve, issued March 9, 1993.
2. Accuntius J, Wilde D. US Patent No 5,284,676, Deposition of pyrolytic carbon in fluid bed, issued February 8, 1994.
3. Ely JL, Haubold AD, Bokros JC, et al. US Patent No 5,514,410, Pyrocarbon and process for depositing pyrocarbon coatings, issued May 7, 1996.
4. Ely J, Emken M, Accuntius J, et al. Pure pyrolytic carbon; preparation and properties of a new material, On-X® carbon, for mechanical heart valve prostheses. J. Heart Valve Dis. 1998;7:626–32.

Chapter 19
Medical Carbon Research Institute LLC/ On-X Life Technologies Inc.

In 1994, Medical Carbon Research Institute was formed to capitalize on the innovations in process control that led to the identification of a superior isotropic carbon called On-X carbon.

These innovations enabled the design of a valve that we hoped would require only platelet inhibitors to control clotting for patients whose only risk factor was the valve itself. If a valve could be designed that preserved blood flow normality, normal hemodynamics, and normal blood trauma, it would have true lifetime durability—a valve for life.

When Medical Carbon Research Institute was formed, Dr. Chambers (from Intermedics, Inc. and Hemex) suggested two surgeons—Professor Ernst Wolner, MD, from Vienna, Austria, and Francis Fontan, MD, from Bordeaux, France—as consultants.

Fontan was a member of the founding committee of the European Association for Cardio-Thoracic Surgery (EACTS), and immediately after its founding, the members unanimously appointed Fontan to be its first president.

Wolner was also involved in the founding of EACTS and hosted its first meeting in Vienna, becoming the 11th president of the organization. Wolner had already established his facility (at the University Hospital in Vienna) as a center of excellence with respect to testing new implant devices, including heart valve replacements.

Visit to a Vineyard

Since Chambers and I were considering an investment in Fontan's vineyard, we visited him in Bordeaux. His vineyard was immense. Fontan told me that he didn't want his sons to become surgeons, as the work was too stressful. Instead, his son Eduardo had obtained a degree in vineyard cultivation.

When we met Eduardo Fontan in the winery, he was dressed in high rubber boots and was scrubbing out the vats. He was tired but expressed genuine gratitude for his

J. Bokros, *Heart of Carbon*, https://doi.org/10.1007/978-3-031-17933-4_19

father having provided him the job. The elder Fontan could not have been any prouder of his son.

At dinner, wine dominated the conversation. Small wooden boxes, each with a special substance inside, were passed around. As the boxes went around the table, we were each to take a whiff. I had to admit that I couldn't smell much. Chambers chided me that I obviously had no future as a connoisseur of wine.

The next evening, we dined at a local vineyard bar and grill, where Eduardo gave me another try at sampling wine. He set two glasses in front of me. "Taste these and pick the good one."

You guessed right: I picked the green one, the one not fully aged. The next day, our final one, Fontan took us to "run the rapids in Bordeaux" (Fig. 19.1). The excursion of Fontan and Chambers kayaking the river was rough and frightening, but they survived. We slept well that night and left in the morning.

Fontan kept inviting me to invest in the Sauternes portion of the Fontan vineyard, but I couldn't quite imagine the potential. This was all the more to do with Chambers's objective: making us "filthy rich." I never bought into that as a responsible objective. It sounded too much like Al Capone's obsession.

On another occasion, after a meeting in Fort Lauderdale, Chambers took Wolner (who had done a clinical trial on the Hemex Duromedics valve) and me to the Bahamas, looking for an off-shore site for an assembly facility in Freeport. After a tour of the island, we returned to Fort Lauderdale (Fig. 19.2).

Fig. 19.1 Professor Fontan and Chambers ready to "shoot the rapids". (Used with the permission of Jack Bokros)

Fig. 19.2 The group lands at Fort Lauderdale. Left to right: Dr. Ernst Wolner; Dr. Russell Chambers; the pilot; and Jack Bokros. (Used with the permission of Jack Bokros)

Fontan and Wolner were invaluable to us. The one fly in the ointment was Wolner's perpetual insistence that our valve was too long. My response continued to be that "the others are all too short."

The 1992 improvement of the carbon process delivered precise control of the processing parameters to produce pure isotropic carbon with markedly improved properties. These processing innovations enabled a new valve design that preserved normal flow of blood through the heart.

Figure 19.3a is a comparison of the respective surface qualities of On-X carbon versus Pyrolite Carbon. Data comparing the flexural strength and fracture toughness for On-X and Pyrolite Carbon are provided in Fig. 19.3b.

First-generation bileaflet valves had been made with Pyrolite Carbon, whose inferior properties imposed manufacturing limitations that restricted design possibilities. A geometrically hemodynamic orifice cannot be achieved with Pyrolite Carbon.

The two advances in technology led to paradigm shifts in processing, making it possible to design a valve that preserves normal flow without turbulence: the long-sought Holy Grail of heart valve replacements.

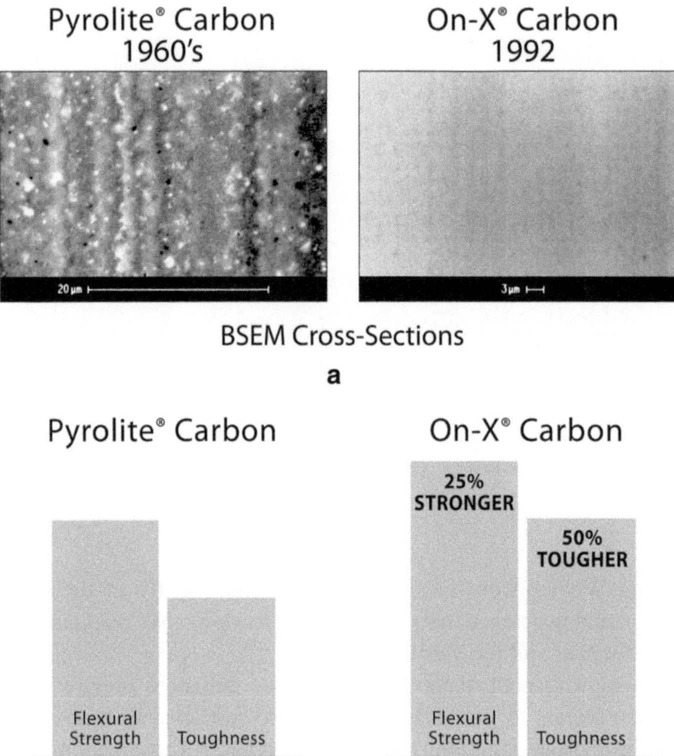

Fig. 19.3 (**a**) Comparison of surface quality and properties of Pyrolite Carbon and On-X carbon (Reproduced with the permission of Jim Kaae). (**b**) Comparison of flexural strength and toughness of Pyrolite Carbon and On-X carbon. (Reprinted from the *Journal of Heart Valve Disease*. Pure pyrolytic carbon: preparation and properties of a new material, On-X® carbon for mechanical heart valve prostheses, JL Ely, et al., 1998 7;6 with the permission of ICR Publishers Ltd)

Chapter 20
Valve Design

In May of 1994, the newly leased Medical Carbon Research Institute building was empty. A facility had to be installed and validated to produce On-X valves. Early every morning, before we engaged in setting up the facility equipment, an hour-long session focused on designing the On-X valve.

The installation of the carbon facility to produce On-X valves was formidable. The image in Fig. 20.1 illustrates the complexity of the process. Such a facility can't be purchased. It must all be customized, including the instrumentation array that controls the flow of hydrocarbon gas in and out of the furnace, as well as the system to maintain the precise temperature of the fluidized bed furnace.

The carbon process evolved in three steps: (1) the process used originally for coating nuclear fuel particles; (2) an imprecise process for coating heart valve components with Pyrolite Carbon; and (3) an optimized process for coating valve components with On-X carbon. This engineering evolution, spanning four decades, was spearheaded by technical professionals with skill sets unique to the endeavor. Visionary personnel like these simply cannot be hired off the street.

Twenty-two seasoned persons had transferred from Carbon Implants, Inc. to Medical Carbon Research Institute, together with ten newcomers added in 1994, creating a facility with finishing and quality programs that were complete and functional. The group is depicted in Fig. 20.2. Just as important, durability testing machines that were capable of accumulating the 600 million cycles required for FDA approvals were functioning by 1995.

As mentioned above, seven persons (Bokros, Ely, Emken, Haubold, Peters, Stupka, and Waits) were involved in the hour-long morning design sessions for the On-X valve. A patent for the design was filed on May 16, 1995, and issued August 1996 [1], with continuations-in-part issued in 1997 [2], 1998 [3], and 1999 [4].

Why the formality of a scheduled meeting every morning? In a word: Medtronic, Inc. When it purchased Carbon Implants, Inc., Medtronic, Inc. imposed a condition that if Bokros were to file for a patent on a heart valve prosthesis in the year following the closing of the purchase of Carbon Implants, Inc., then Medtronic, Inc. would own that patent.

J. Bokros, *Heart of Carbon*, https://doi.org/10.1007/978-3-031-17933-4_20

Fig. 20.1 Flow system used to deposit On-X carbon on heart valve replacement components. The two white cylinders surrounded by copper coils are induction-heated fluidized bed reactors. (Used with the permission of Jack Bokros)

However, if a valve patent could be shown to be developed independently in the first year, Medtronic, Inc. would then have no right to it.

Each meeting was documented by Jonathan Stupka, Head of Operations for Medical Carbon Research Institute. Each and every feature thought to be patentable was recorded along with the person initiating the idea.

The first question addressed was to determine if the length-to-diameter ratio of the valve when open is important. In the past, the adage was that the "profile" should be as low as possible—a myth perpetuated by implanting surgeons. We asked Ned Hwang to study flow through tubes with various length-to-diameter ratios at the flows experienced during systole. His results appear in Fig. 20.3. The ordinate represents the pressure loss as a function of its length portrayed by the length-to-diameter ratio depicted on the abscissa. Each plot is for individual flow rates that range from 10 to 30 liters/minute, the range experienced during systole.

As the length-to-diameter ratio increases, the pressure loss across the tubes drops rapidly and levels out at a L/D ratio of about 0.6. The value chosen, 0.62, is close to the 0.7 L/D value reported by Swanson and Clark for the native aortic valve (Fig. 20.4) [5]. The length is a bit shorter than 0.7 to placate those influenced by the myth that the valve should be as short as possible. We defied the myth with a ratio

Fig. 20.2 Twenty-two employees transferred from Carbon Implants, Inc. when it was purchased by Medtronic, Inc. There were a number of additional new hires to fill out the slate. (Used with the permission of Jack Bokros)

of 0.62—no less! This L/D ratio is two to three times that of other valves. The new length could substantially reduce the risk of pannus intrusion on the inlet and outflow sides of the valve.

The data relating to L/D in Figs. 20.3 and 20.4 in the previous paragraphs refute the low-profile myth.

When I showed Endre Bodnar Hwang's result, he cocked his head thinking. After a pause he said "Bokros is Hungarian, right?" He knew we had common heritage. "The ancient Egyptians knew that lengthening a tube would enhance flow 4000 years ago! Go to the library in Budapest and look up Pattantyus, G. Gyakorlati: áramalastan, Tankönyvkiado, Budapest, 1959."

Kathleen Selbrede reacted and produced Fig. 20.5a–c.

A decade earlier in Budapest, I presented a paper mentioning two valves for China. Endre was the chairman of the meeting. Following my presentation, he commented with a smile, "you remind me of *The Pilgrim's Progress*." I had not read *Pilgrim's Progress* and shrugged it off. Later, I did learn that the book is a Christian allegory and the main theme is about perseverance in the face of trials and tribulations.

At the time, Endre Bodnar was the founder and editor of the *Journal of Heart Valve Disease*. We will see more from Endre later.

The prevalence of the low-profile myth was reinforced at that time when a prominent surgeon announced the Wada-Cutter valve—with the lowest profile ever—as a

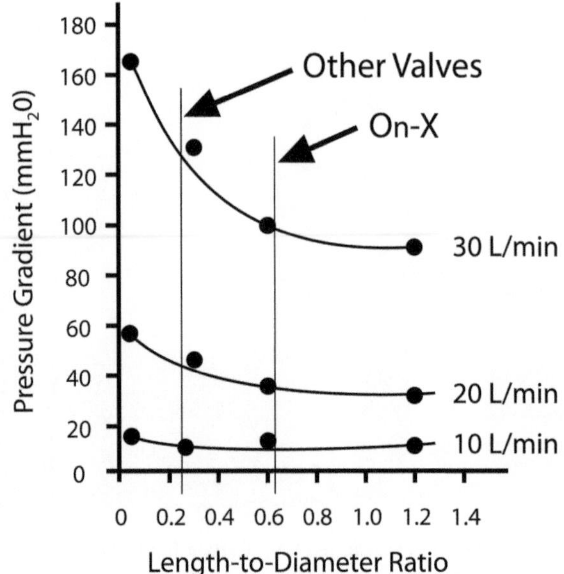

Fig. 20.3 Pressure gradient (mm H_2O) versus the length-to-diameter ratio of tubes for flows in the range experienced during systole (10–30 liters per minute). (Reproduced with the permission of Ned HC Hwang)

design breakthrough. He extolled the Wada's low profile as justification for an artificial heart design incorporating four Wada valves.

A Flare for Valve Design

It was obvious that the valve inlet had to be flared to eliminate the flow separation that forms a turbulent area (the vena contracta). In 1990, before On-X carbon was developed, Helmet Reul of the Helmholtz Institute of Aachen, Germany, an authority on heart valve fluid dynamics, presented a design that had a flared inlet, asking if we could make it.

I replied "Sure." But I told him that while such a valve *could* be made, the leaflets could not be inserted by stretching the orifice without having it fracture in the process.

This was so because Pyrolite Carbon did not have the fracture strength to allow the leaflets to be inserted by stretching. In any case, Helmet's design was too short—a shortcoming we've already elucidated. It was very likely that Helmet had been influenced by industry scuttlebutt circulating among surgeons.

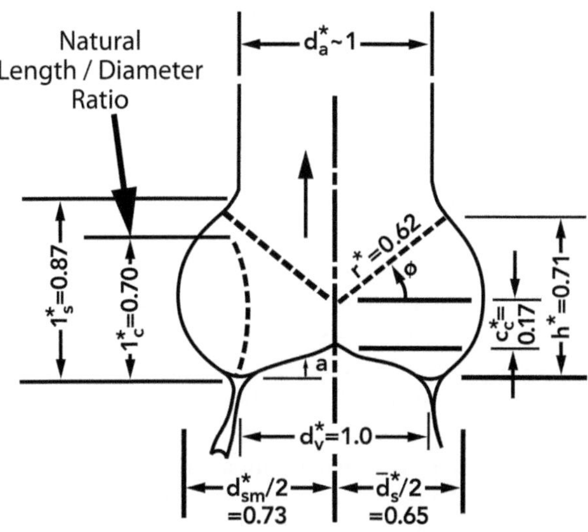

Fig. 20.4 Data reported by Swanson and Clark showing the height-to-diameter ratio for the native human aortic heart valve when open. (Reprinted from *Circulation Research.* Dimensions and geometric relationships of the human aortic value as a function of pressure, W Milton Swanson, RE Clark, 1974;35(6) with permission from Wolters Kluwer Health, Inc.)

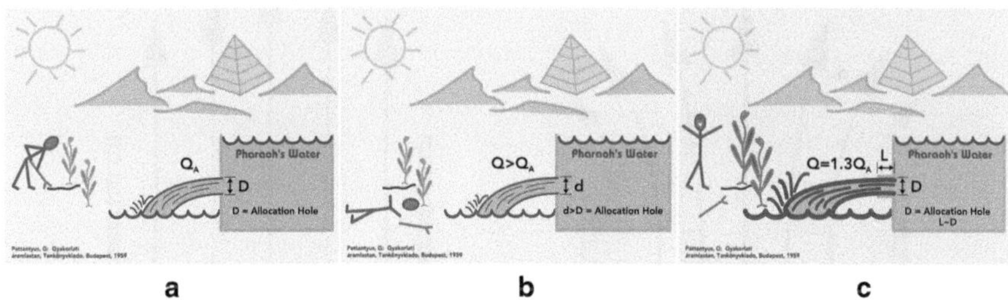

Fig. 20.5 (**a**) Four thousand years ago, Pharaoh owned all the water; (**b**) if anyone tampered with the allocation port, he was severely punished; (**c**) an enterprising peasant discovered that if he added a tube at the water exit with a diameter the same as the allocation hole and with the length equal to its diameter, his flow increased by 1.3. (Used with the permission of Artivion, Inc. and Kathleen Selbrede)

Professor Hwang determined the proper radius of curvature of the flare and its correct positioning by testing various flared graphite tubes. The flare had to be large enough to avoid flow separation but small enough so as not to incur an implantation problem. The textbook value of the radius of curvature needed to be equal to or greater than one-seventh of the diameter of the valve inlet.

The flare was positioned intra-annularly; the sewing cuff was positioned supra-annularly, with the flare extending inter-annularly.

Professor Hwang had intimate familiarity with all aspects of hemodynamics. Everyone remembers his most oft-repeated axiom: "Don't fight the flow. Go with the flow; follow the flow." Fighting the flow leads to turbulence. Another of our advisors, Walter Dembitski, MD, San Diego, California, put it this way: "Turbulence makes the blood angry—it reacts by clotting." [6].

Flow separation at the inlet of a straight cylinder (Fig. 20.6a) causes turbulence downstream of the leading edge, directly reducing the diameter of the organized flow (this reduction is the "vena contracta"). If the inlet is flared (Fig. 20.6b), flow will not separate so long as the radius of curvature of the flare is greater than one-seventh of the diameter. The flow rate depends on the fourth power of the diameter, so the resulting increase will be substantial.

Figure 20.7 shows a schematic comparison of the flow through three different structures: a conventional bileaflet heart valve replacement, an On-X aortic valve replacement, and a native valve. The leaflets of a current valve on the left stop abruptly before they reach the full open position. This generates turbulence. If, however, the inlet is flared (like the natural valve shown in the center location and the On-X valve on the right), turbulence can be eliminated.

Like the natural valve, the On-X valve's leaflets are free to follow the flow to full open, parallel with the flow. This was accomplished with an actuated purgeable pivot design (Fig. 20.8a). The leaflets fit into the precisely engineered pivot sockets using specially designed leaflet tabs. The pivot sockets incorporate closing ramps

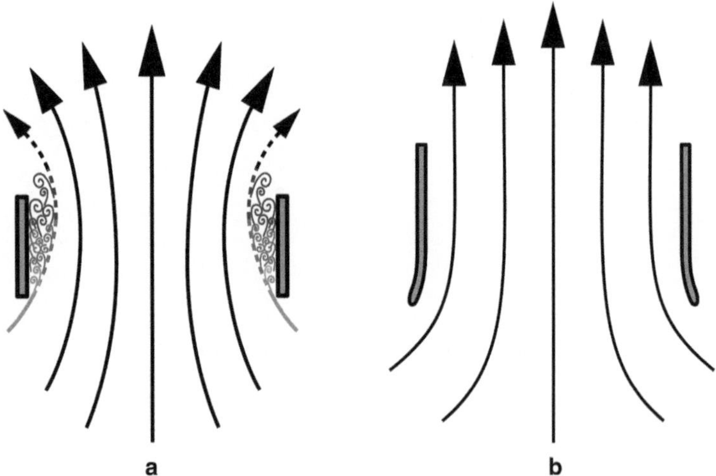

Fig. 20.6 (a) Flow separation occurs when fluid flows into a nozzle. Flow separation at the inlet produces a circumferential region of turbulence that blocks the organized flow and reduces its diameter. This reduction is called the "vena contracta." (b) When the inlet is flared, flow separation at the inlet and the vena contracta are eliminated, and the flow diameter increases. To be effective, the flare's radius of curvature must be at least about one-seventh of the nozzle's inside diameter. (Used with the permission of Artivion, Inc. and Kathleen Selbrede)

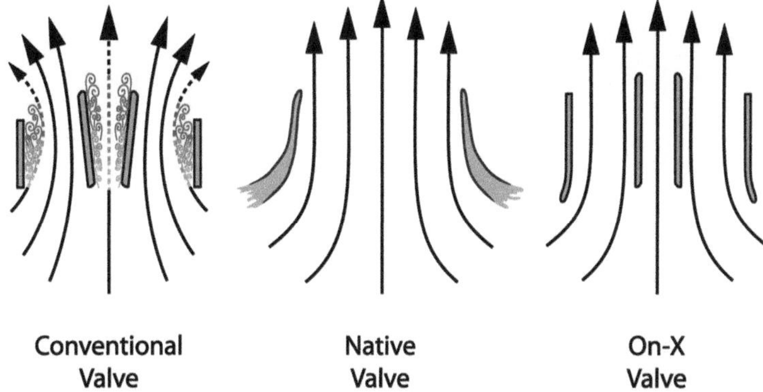

Fig. 20.7 Flow through a conventional bileaflet valve is turbulent because the orifice is straight. The inlet lacks a flare and the leaflets are stopped before they can open fully. Flow through the On-X valve is not turbulent because its design mimics the native valve: the orifice is long, the inlet is flared, and the leaflets open fully. (Used with the permission of Artivion, Inc. and Kathleen Selbrede)

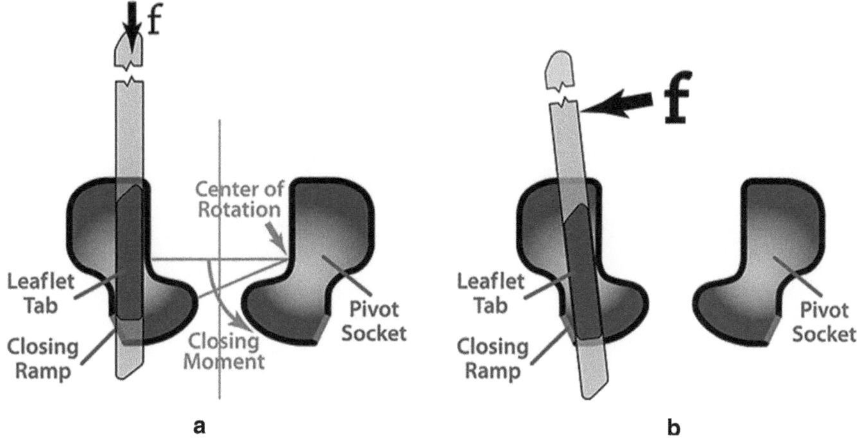

Fig. 20.8 (**a**) Actuated pivot. (**b**) Closing ramps. (Used with the permission of Artivion, Inc. and Kathleen Selbrede)

that come into play when blood flow reverses (Fig. 20.8b). Reverse blood flow exerts force, f, on the leaflets, causing the leaflet tabs to slide down the closing ramps. This "actuation" reliably positions the leaflet at the threshold angle to guarantee closure. Flat, thin, On-X carbon leaflets are unobstructed, free to align themselves with the streamline normal blood flow and close reliably.

The depiction of the On-X valve at late systole (Fig. 20.9a, b) demonstrates streamlined flow separation into two flows. The first is mid-stream and appears straight but has a mild helical flow component. The second flow demonstrates that the outermost portion diffuses into the sinuses, where the flow becomes vortical,

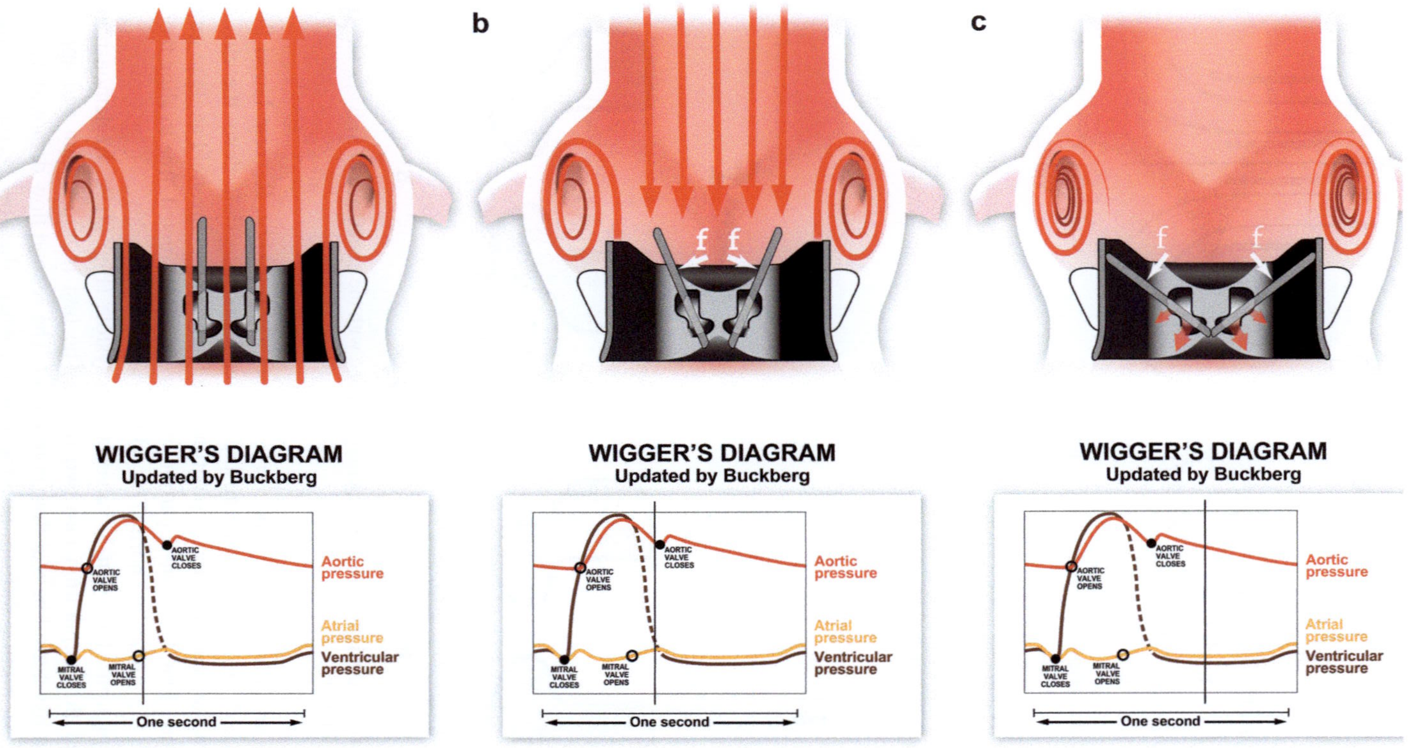

Fig. 20.9 (**a**) A split second before the actuation of the pivot (demonstrated in Fig. 20.8b), the mitral valve begins to open. (**b**) As the aortic valve closes, the reversed flow from the mitral valve closes the aortic valve, and aortic pressure propels spiraling flow into the coronaries. (**c**) Momentum propels the spiraling flow into the coronaries as the pivots are purged by aortic pressure. (Used with the permission of Kathleen Selbrede)

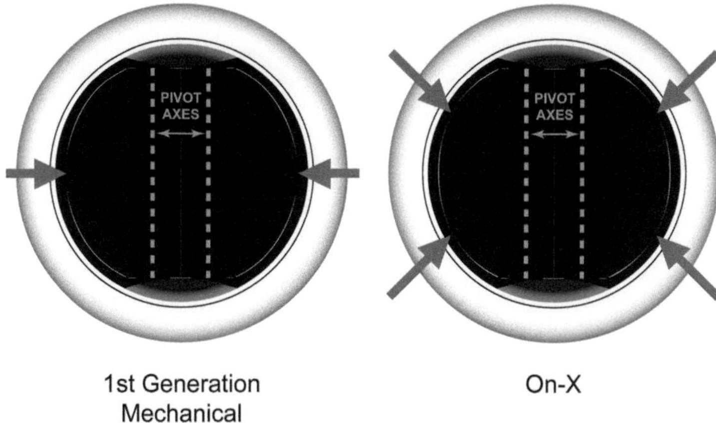

Fig. 20.10 First-generation mechanical valve leaflets contact the orifice with a high-velocity single-point direct hit (left). On-X leaflets land softly on the orifice with a two-point glancing contact (right). (Used with the permission of Kathleen Selbrede)

storing rotational momentum (conservation of rotational momentum drives the helical flow into the coronaries during diastole, as depicted in Fig. 20.9c).

Following the Wigger diagrams (Fig. 20.9a–c), observe that the mitral valve opens before the aortic valve closes, as reported by Buckberg.

High-Impact Versus Low-Impact Valve Closure

Compared with first-generation mechanical valves, the closure of the On-X valve is low-velocity and low-impact, similar to the slow, gentle closure of a healthy native valve.

First-generation mechanical valve leaflets make hard contact with the orifice in a high-velocity direct hit at their tips (Fig. 20.10 left). On-X leaflet tips (Fig. 20.10 right) never contact the orifice. This is because the curved outer rim of the leaflets is shaped so they contact the orifice at only two points offset 45° from the tip. The On-X leaflet makes a softer, glancing (oblique) contact with the orifice at two points, leaving a tiny gap from one contact point all the way past the tip to the other contact point. This design reduces the On-X leaflet's force of impact. Unlike earlier designs, On-X simply don't slam shut in an injurious way.

As On-X leaflets close, depicted in Fig. 20.11, their rotational velocity, compared with first-generation mechanical valve leaflets, is reduced because the radial distance from each contact point to the axis of rotation is reduced.

Impact forces, depicted in Fig. 20.11, that are perpendicular to the orifice wall on the right are distributed between two contact points, reducing the impact at each point. Rectilinear vector analysis of the primary tangentially directed velocity vector shows a substantial reduction in the impact provided by the 45° offset. The resulting soft landing prevents cavitation and blood damage. Experimentally

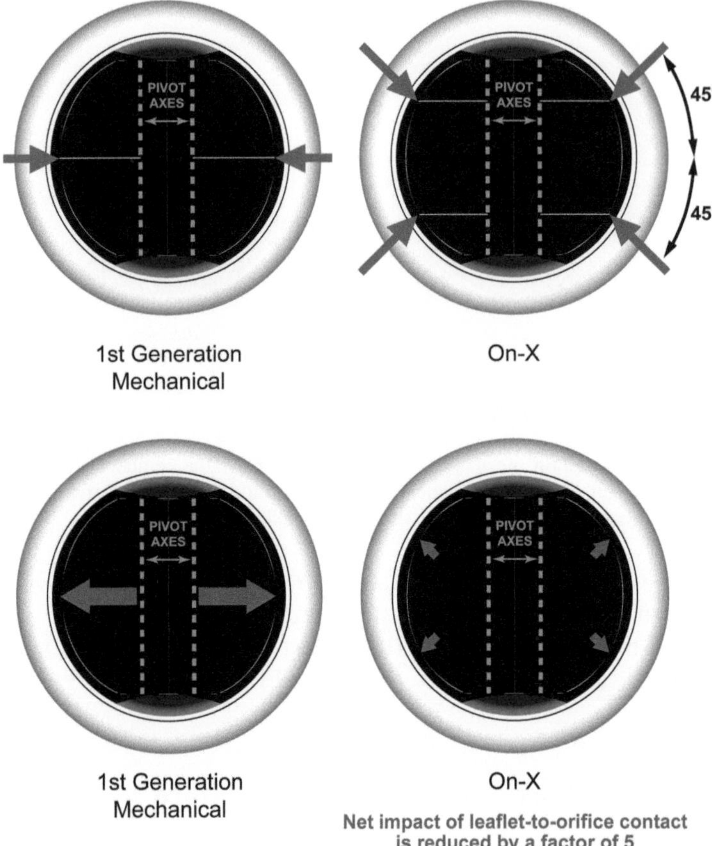

Fig. 20.11 For first-generation mechanical and On-X valve designs, the net impact of leaflet-to-orifice contact is substantially reduced. (Used with the permission of Kathleen Selbrede)

measured regional backflow velocity transients provided by Larry Scotten's instrument described in coming paragraphs are convincing.

Two distinct backflows between the leaflet contact points with the orifice wall create a hydraulic cushion that mitigates the "water hammer" effect, preserving red blood cell (RBC) integrity and reducing hemolysis. The backflows through the pivots keep them clear by eliminating stasis (Fig. 20.12). The energy expenditure invested in these two backflows is very small when measured against their tremendous benefits. Evidence reveals that the On-X soft landing design minimizes noise and cavitation potential, as well as hemolysis.

Experimentally measured regional backflow velocity transients (closing impact) for the On-X aortic valve (Fig. 20.13) are substantially lower during leaflet closure than the same transients measured in the St. Jude Medical Regent valve. The results were measured using a precision electro-optical instrument (appropriately named "The Leonardo") developed by Larry Scotten's ViVitro Systems, Inc. [7]. Mitral

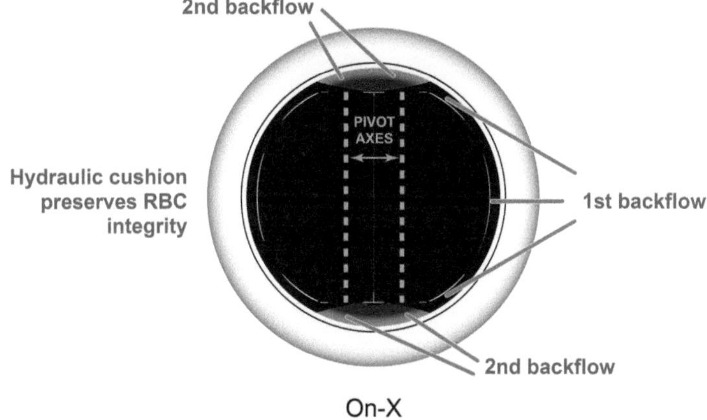

Fig. 20.12 Backflows that occur on closure of the On-X valve create a hydraulic cushion for the leaflets and also purge the pivots. (Used with the permission of Kathleen Selbrede)

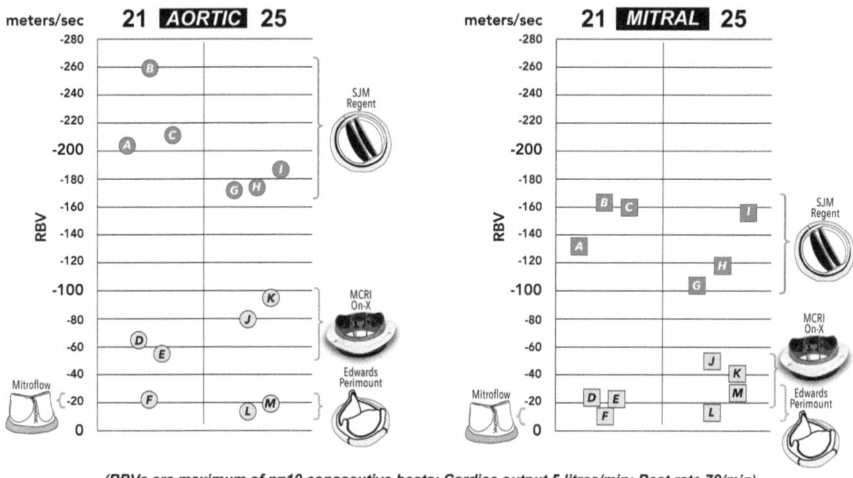

Fig. 20.13 Measurements of regional backflow velocity (RBV) transients (closing impact) at or near valve closure confirm that On-X leaflet closure is gentle; values for On-X are closer to those of tissue valves than to those of first-generation bileaflet valves. (Courtesy of LN Scotten per ViVitro Systems, Inc.)

valves (on the right) have lower closing impact values than aortic valves because their leaflets are propelled by the left ventricle's systolic contraction. This is why the On-X soft landing closure is more pronounced in the mitral position.

The hemodynamic performance of prototypes, together with their durability and the absence of cavitation, was established so that the valve was ready for its first implant by September of 1996. The final design is depicted in Fig. 20.14. The valve is protected by patents and a number of divisional patent applications [1–4].

Three of the granted applications were for process patents [8–10].

United States Patent

Patent Number: **5,772,694**

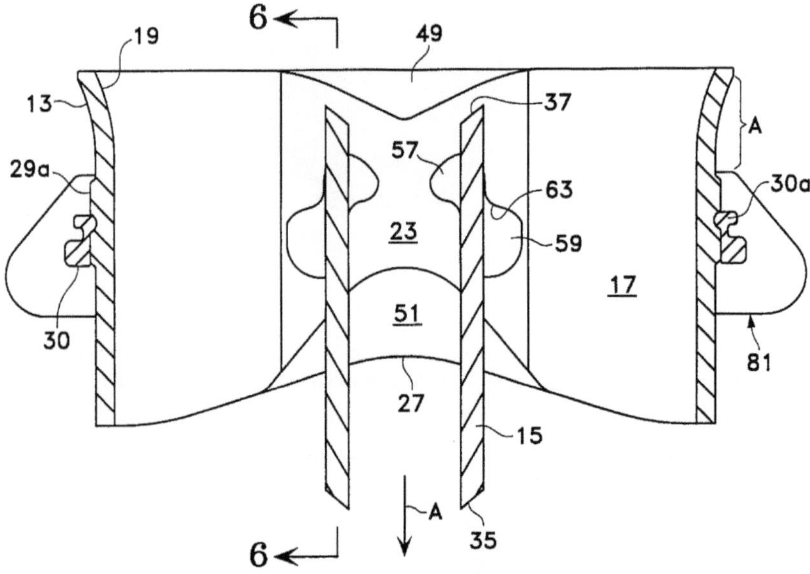

Fig. 20.14 Bokros JC et al. US Patent No 5,772,694, issued June 20, 1998

References

1. Bokros JC, Ely JL, Emken MR, et al. US Patent No 5,545,216, Prosthetic heart valve with improved blood flow, issued August 18, 1996.
2. Bokros JC, Ely JL, Emken MR, et al. US Patent No 5,641,324, Prosthetic heart valve with improved blood flow, issued June 24, 1997.
3. Bokros JC, Ely JL, Emken MR, et al. US Patent No 5,772,694, Prosthetic heart valve with improved blood flow, issued June 30, 1998.
4. Bokros JC, Ely JL, Emken MR, et al. US Patent No 5,908,452, Prosthetic heart valve with improved blood flow, issued June 1, 1999.
5. Swanson WM, Clark RD. Dimensions and geometric relationships of the human aortic valve as a function of pressure. Circ Res. 1974;35:871–82.
6. Dembitski G, Personal communication, November 1995.
7. Scotten LN, Siegel R. Importance of shear in prosthetic valve closure dynamics. J. Heart Valve Dis. 2011;20:664–72.
8. Emken MR, Seeley M, Wilde DS. US Patent No 5,305,554, Moisture control in vibratory mass finishing systems, issued April 26, 1994.
9. Wilde DS, Hightower BF, Accuntius JA. US Patent No 5,332,337, Particle feeding device and method for pyrolytic carbon coaters, issued July 26, 1994.
10. Waits CT, Stupka JC, Peters TS, Schwartz AS. US Patent No 5,336,259, Assembly of heart valve by installing occluder in annular valve body, issued August 9, 1994.

Chapter 21
Clinical Trials

Anticipating a quick approval afforded by the unprecedented unique design of the valve and its use of pure isotropic On-X carbon, we planned auxiliary studies to demonstrate the On-X valve's thromboresistance using protocols consistent with FDA requirements. The purpose was to accelerate the grant of an IDE that would allow for a reduction in anticoagulation requirements.

Medical Carbon Research Institute, under the direction of John Ely, started clinical trials beginning in Europe and extended to the United States using a protocol that was crystal clear with respect to control of international normalized ratio (INR). Ely had already submitted four applications for pre-market approvals. All four were approved.

Prof. Ernst Wolner recommended Professor Axel Laczkowits (Fig. 21.1) at Bochum University to perform the first On-X implant. The implant was done on September 12, 1996 [1].

The CE mark for the On-X valve was awarded in 1998.

In order to maintain a continuing perspective, Figs. 21.2 and 21.3 document important events and goals. Figure 21.2 is a schematic timeline that, while not precisely accurate, provides representative market trends relating to the growth and decline of the mechanical and tissue heart valve replacement segments of the market.

A Pause to Reflect Important Events

In the decade before 1960, the development of heart valve replacements was chaotic and unfocused. Those years were documented in the proceedings of a conference that was held in Chicago in 1960, edited by Merendino [2] and published in 1961. In his closing remarks, the conference chairman, Dr. Merendino, expressed disappointment and frustration: "The road to the development of a prosthetic valve has been rough with potholes; there was not even a perceptible trail forward to be

© The Author(s), under exclusive license to Springer Nature Switzerland AG 2023
J. Bokros, *Heart of Carbon*, https://doi.org/10.1007/978-3-031-17933-4_21

Fig. 21.1 Prof. Axel Laczkowits of Bochum University, Bochum, Germany, implanted the first On-X valve on September 12, 1996. (Used with courtesy of Axel Laczkowits)

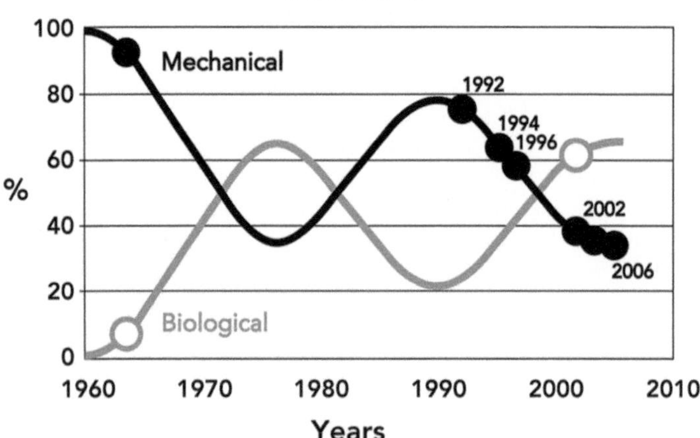

Fig. 21.2 Schematic historical market share between mechanical and tissue heart valve replacements. (Used with the permission of Artivion, Inc.)

found." He forecast that future development would need to be redirected along entirely new lines of inquiry.

The early 1960s saw two important advances in material technology. One was due to Alain and Sophie Carpentier when they developed the glutaraldehyde method of fixing biological tissue. This method provided focus on the biological side [3].

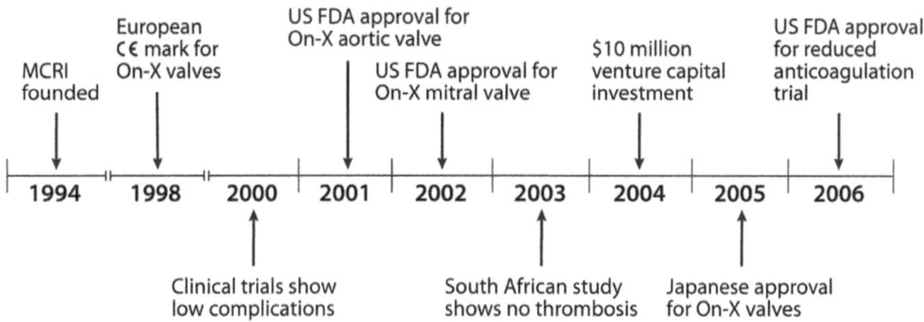

Fig. 21.3 Historical record of Medical Carbon Research Institute/On-X regulatory approval milestones. (Used with the permission of Artivion, Inc.)

The other advance occurred in 1963, while I was working on an unrelated project at the General Atomic facility in San Diego and discovered the isotropic form of pyrolytic carbon. That discovery provided focus on the mechanical side [4].

The relative trends in Fig. 21.2 show a rapid penetration of the market by the introduction of the Hancock biological prosthesis up until about 1978. The first reports of failures arose from the Mayo Clinic.

The decline in biological prostheses starting around 1979 led to another peak in the mechanical sector, when the ratio of mechanical to tissue valves rose nearly four to one in favor of mechanical variants.

In 1990, Edwards introduced the Carpentier series of biological valves, setting in motion the rapid recapture of market share from the mechanical segment.

On Fig. 21.2, the year 1992 marks the point where Accuntius and Wilde filed patents on a new control apparatus for the deposition of pyrolytic carbon in a fluidized bed [5] (Fig. 18.10 in Chap. 18). This advance made possible the discovery of On-X carbon by Ely and his colleagues in 1996 [6] with the results being published in 1998 [7].

The year 1994 (Fig. 21.2) marks the date that the Medical Carbon Research Institute was formed. Referring again to Fig. 21.2, when Medical Carbon Research Institute was formed, note that the mechanical side of the market was falling rapidly, with the market share moving decisively toward the tissue side. This was a testament to the effectiveness of the Edwards marketing initiative. Edwards vigorously promoted the durability of their Carpentier valve line while highlighting the continuing problems associated with anticoagulation complications caused by Coumadin in competing mechanical valves.

The design and testing of the On-X valve also moved rapidly, starting in 1994 (Fig. 21.3) and continuing from conception to the gaining of the CE mark in 1998, with clinical trials showing very low thrombotic and bleeding complications.

In 2000, as Medical Carbon Research Institute was approaching pre-market approval, it needed sales/marketing experience. This was the same situation we had faced in 1985, when CarboMedics, Inc. hired Rollie Siegel. After the European company purchased Intermedics, Inc./CarboMedics, Inc., the management team at

CarboMedics, Inc. left to form Carbon Implants, Inc. (Chap. 18) and beyond Carbon Implants, Inc. to form Medical Carbon Research Institute leaving Siegel at CarboMedics, Inc. Déjà vu. We needed Siegel again.

I contacted Rollie and brought him up to date. He was interested but hesitant, as he did not want to be portrayed as a job hopper. He had to think about it—maybe take a leave of absence to consider retirement. He'd have to get back to me.

Get back he did, and as was his custom, he was retained as a consultant with an annual fee and an incentive package.

In 1999, John Ely and I met with Mervyn Williams, MD (Fig. 21.4), of Port Elizabeth, South Africa, at the Society for Heart Valve Disease in London. Mervyn especially liked the long orifice, which was nothing like valves preceding On-X that still used (in his words) a "washer-like" orifice with virtually no consideration of the threat of pannus encroachment into the valve. Shortly after our meeting with Williams, Ely set up a clinical trial in South Africa.

The orderly progression displayed in Fig. 21.3 was no accident. This is because in the years between 1990 and 1994, the same team had developed a bileaflet valve for Medtronic, Inc., as well as a facility with sufficient capacity to produce 40,000 of these valves per year.

Although the technical progression was on schedule, and the On-X valve was showing promise as a veritable dream valve, there were storm clouds dead ahead. They were the harbinger of a perfect storm heading our way.

At the inception of Medical Carbon Research Institute, we had anticipated that as approvals were granted, there would be a steady stream of capital from the royalty on Medtronic, Inc. Parallel Valve sales. But the Medtronic, Inc. clinical trial failed (Chap. 18). In addition, the fall in market share for mechanical valves was affecting our bottom line. The rapid upsurge on the tissue side of the industry

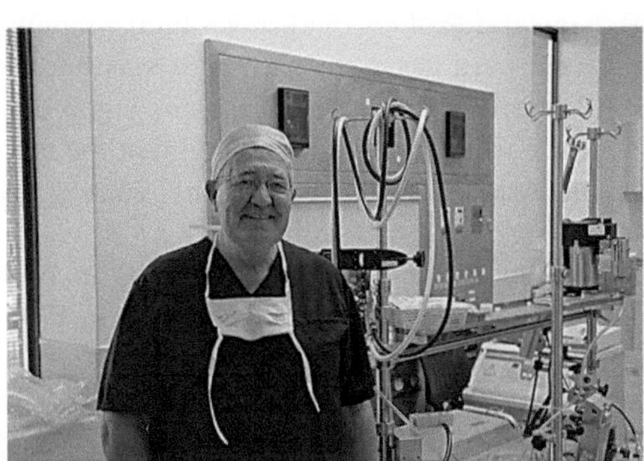

Fig. 21.4 Mervyn and Bernadette Williams. (Used with the permission of Mervyn and Bernadette Williams)

(Fig. 21.2) and the failure of Medtronic, Inc.'s Parallel Valve clinical trial were factors.

Referring back to Chap. 18, the details of the Medtronic, Inc. trial were hidden, protected by a confidentiality agreement. The reality behind the reported clotting was never publicly reported. The failure of the Parallel Valve was startling. It directly contradicted the uniformly excellent pre-clinical data displayed in the Design Freeze report in Chap. 18. The Design Freeze report indicated that the Parallel Valve's performance was better than any other valve, mechanical or biological.

How then could the Medtronic, Inc. clinical trial have failed? Endre Bodnar addressed the unexplained thrombosis of the Medtronic, Inc. Parallel Valve in the *Journal of Heart Valve Disease*, 1996; 5: 572–3, "The Medtronic Parallel Valve and Lessons Learned." Bodnar was eminently responsible and thorough in his analysis. Lacking any explanation, he naturally assumed that any trial in humans would follow professional norms.

In his sixth paragraph, he concluded that "the failures could not be forecast by the knowledge available at the time." I am relieved that someone shed light on the Parallel episode exposing the disruption of scientific thought.

As yet there has been no response to Endre's learned editorial.

When a prominent surgeon was approached to do a clinical trial on the On-X valve, he asked me, "Aren't you the guys that made the Parallel?"

Boom. End of conversation.

Investment bankers issued similar responses to us. Medical Carbon Research Institute was "high and dry," its funds approaching zero. In the midst of this crisis, we were granted FDA approvals in 2001 and 2002 for our aortic and mitral valves, respectively [8, 9]. Despite meeting these important milestones, by 2002, we were literally up against the wall. We weren't completely without options, however:

1. We could declare bankruptcy, but that option was not the best, as the On-X valve would be viewed as an orphan valve, i.e., one without a home. History has shown that such valves are not marketable, so bankruptcy really wasn't an option.
2. The other option wasn't much better. We could strike a deal with a venture capitalist. They would then lay down the law to us: "This is all we'll risk and we'll only risk it on our terms."

We took option 2, which preserved the technology and (we hoped) the success of the project. In any event, only one venture capital group expressed interest in us.

We made a deal, starting in 2002, with a venture capital group that declared they were not to be categorized alongside venture capitalists labeled "Vulture Capitalists." Their offer for Class A shares that sold for $4,000 per unit were valued at 3 cents.

They were the only game in town and so that was it.

The valve was already approved for sale in the United States and Europe, yet somehow it was now worth only 0.00075% of each dollar that had been invested to create it?

At least Medical Carbon Research Institute's stock wasn't valued at zero. The company was still intact, and the best valve on the market was still available to patients.

Fig. 21.5 (**a, b**) PowerPoint title slides used by Rollie and Jack. (Used with the permission of Artivion, Inc)

When Siegel came aboard, he and I teamed up for valve presentations in the same way we had done before at CarboMedics, Inc., preferring to structure these as dinner presentations. I would lead with technical material, although Siegel was the true lead since he was the door opener. After my discussion was concluded, he would pick up the clinical part. Once in front of a group, either one of us could present either portion of the talk (Fig. 21.5).

Our motivation in designing the On-X valve was to eliminate the need for anticoagulation. To achieve that goal, the valve had to incur no more blood damage than the native valve did. In that light, we focused on minimizing the valve's propensity to produce turbulence. If we could produce a valve with flow that mimicked the native valve, the consequences would be natural hemodynamics, normal blood trauma, and a reduced need for anticoagulation, i.e., normality.

References

1. Laczkovics A, Heidt M, Oelert H, et al. Early experience with the On-X prosthetic heart valve. J Heart Valve Dis. 2001;10:94–9.
2. Merendino AK (ed). Proceedings of conference on prosthetic heart valves for cardiac surgery, September 9–10, 1960. Springfield: Charles Thomas; 1961.
3. Carpentier A, Lemaigre G, Robert L, et al. Biological factors affecting long-term results of valvular heterografts. J Thorac Cardiovasc Surg. 1969;58:457–83.
4. Bokros JC. Random pyrolytic carbon. Nature. 1964;202:1004.
5. Accuntius J, Wilde D. US Patent No 5,284,676, Deposition of pyrolytic carbon in fluid bed, issued February 8, 1994.
6. Ely JL, Haubold AD, Bokros JC, et al. US Patent No 5,514,410, Pyrocarbon and process for depositing pyrocarbon coatings, issued May 7, 1996.

7. Ely J, Emken M, Accuntius J, et al. Pure pyrolytic carbon; preparation and properties of a new material, On-X® carbon, for mechanical heart valve prostheses. J Heart Valve Dis. 1998;7:626–32.
8. Summary of safety and effectiveness for On-X 2001 aortic, FDA PMA P000037, May 30, 2001 and March 6, 2002 European Primary Trial Updated May 31, 2003.
9. Summary of safety and effectiveness for On-X 2002 mitral, FDA PMA P000037/S1, May 30, 2001 and March 6, 2002 European Primary Trial Updated May 31, 2003.

Chapter 22
Design Validation

Pannus tissue that attaches to the sewing cuffs of both mechanical and tissue valves is an inevitable threat to all heart valve replacements. The On-X design inherently protects against pannus encroachment. Left unchecked, pannus can cause valve failure. In biological valves, fixed biological tissue reacts with pannus to accelerate its encroachment, per Peter Zilla [1].

Traditional mechanical valves' short blunt orifice and (in the case of the St. Jude Medical valve) the close proximity of pivot guards to the septum render them vulnerable to pannus complications.

The On-X valve's long flared orifice (Fig. 22.1) shelters the leaflets on both the inlet and outflow sides in any orientation. Pannus stops when it encounters carbon, because it does not recognize carbon as a foreign body. No pannus encroachment has been reported in the 200,000 On-X valve implants performed over 19 years—not even in the 4000 South African patients (through November 2014) who are inconsistent in taking their prescribed anticoagulation medication.

Blood damage was determined by monitoring lactate dehydrogenase (LDH) (Fig. 22.2). The data compares LDH levels for On-X patients with data from the other two bileaflet valves. On-X patients, both aortic and mitral, stayed solidly in the normal range, while LDH levels for the other two valves were elevated.

The hemodynamic data were revealing. Values for mean and peak gradients (Fig. 22.3) showed the On-X valve gradients were in the normal range and lower than other bileaflet valves. The corresponding effective orifice area and the values indexed to body surface area (Fig. 22.4a, b) show no indication of patient-prosthesis mismatch.

Our design objective was to eliminate turbulence. At that time, with normal versus abnormal blood flow not yet fully defined, we followed our intuition that normal blood flow would be a direct consequence of normal anatomical geometry.

© The Author(s), under exclusive license to Springer Nature Switzerland AG 2023
J. Bokros, *Heart of Carbon*, https://doi.org/10.1007/978-3-031-17933-4_22

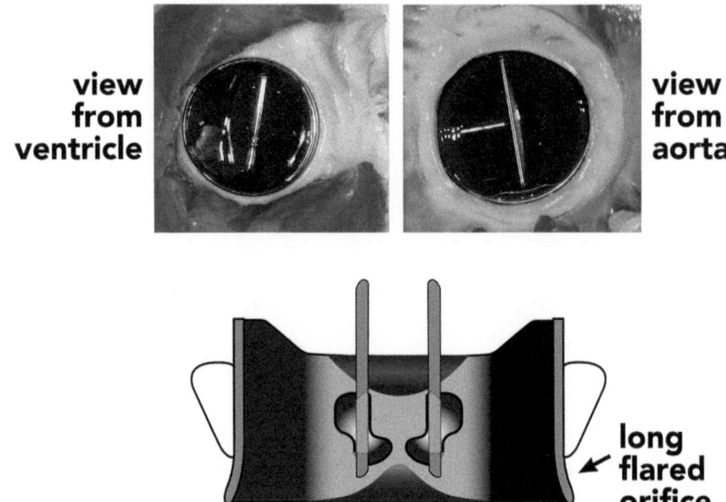

Fig. 22.1 On-X long flared orifice protects from pannus encroachment. (Used with the permission of Artivion, Inc.)

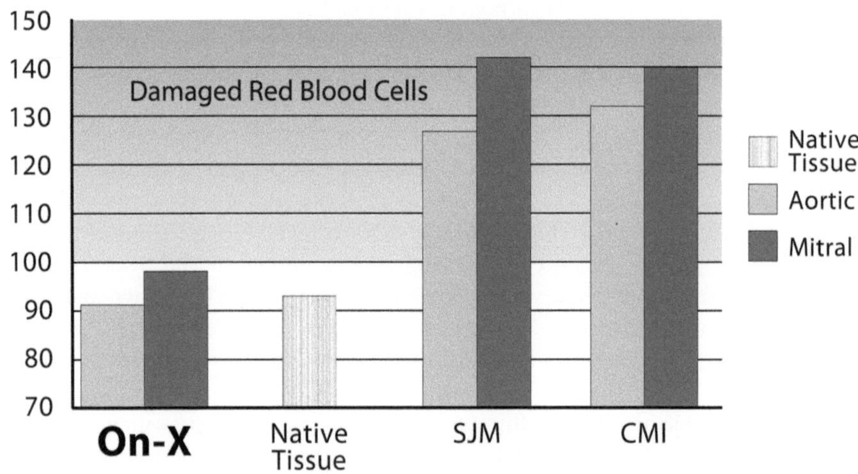

Fig. 22.2 Blood damage stays in the normal range for the On-X valve as shown by measurements of lactate dehydrogenase (LDH) released by damaged red cells. (Used with the permission of Artivion, Inc.)

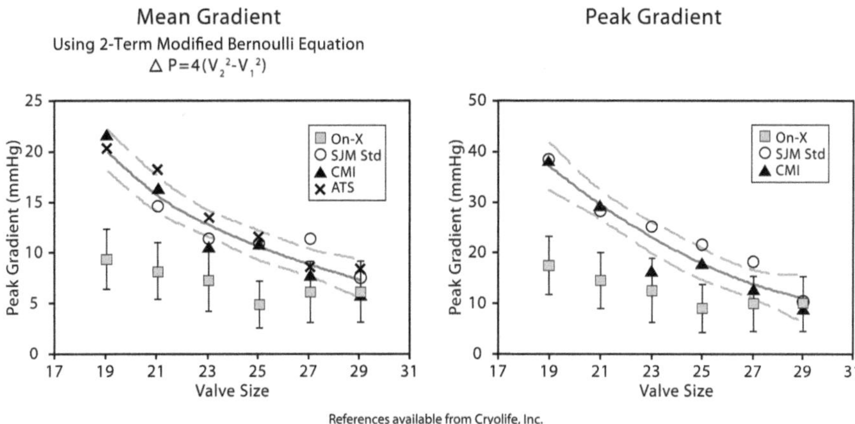

Fig. 22.3 Mean and peak pressure gradient values for the On-X valve coincide with those of the native valve [2]. (Used with the permission of Artivion, Inc.)

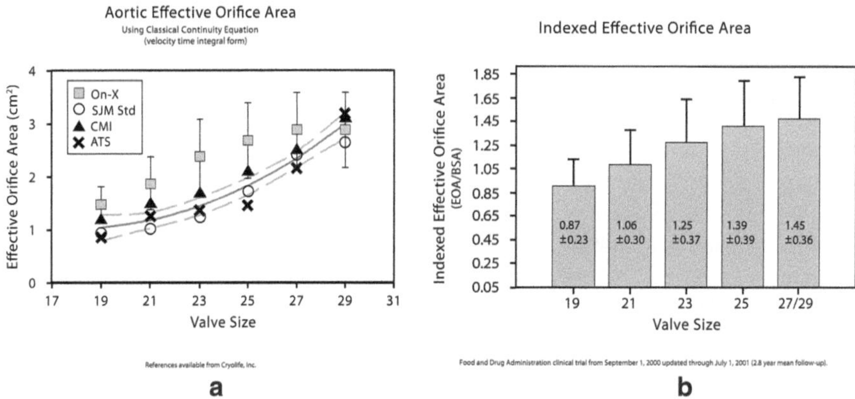

Fig. 22.4 (a) Effective orifice areas and (b) indexed body surface area [2] (Used with the permission of Artivion, Inc.)

Leonard da Vinci Lays Down the Law

One of the first clues regarding blood flow normality appeared in the work of Leonardo da Vinci. His remarkably detailed laboratory notes [3, 4] provide a precise description of intra-sinus vortical flow in the heart of an ox: "Blood flows out of the ventricle into the sinuses of Valsalva, where a portion of the forward flow is diverted to form a vortical flow that assists in the closure of the leaflets of the aortic valve." After the leaflets have closed, the residual vortex is directed with a lateral motion toward the coronaries, developing into a spiraling, or helical, flow that propels blood into the coronaries.

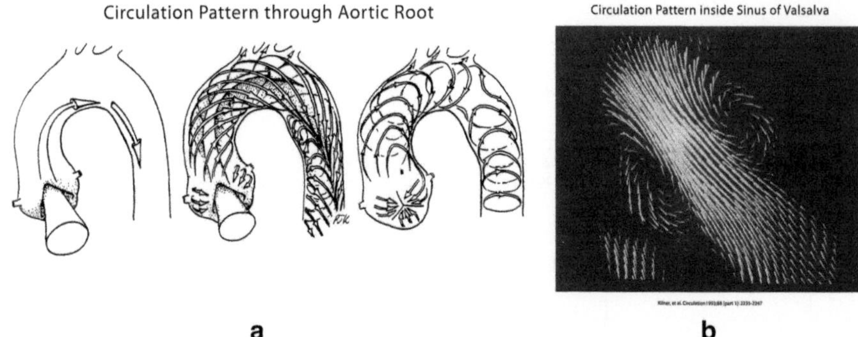

Fig. 22.5 (**a**) Kilner observed that blood ejected from healthy human hearts flows in a spiral pattern through the aortic root (**b**) and then separates in the sinuses to form a vortical flow that assists leaflet closure. (Reprinted from *Circulation*, Helical and retrograde secondary flow patterns in the aortic arch studied by three-directional magnetic resonance velocity mapping, PJ Kilner, et al., 1993;88(5) with permission from Wolters Kluwer Health, Inc.)

In the 1990s, Philip Kilner et al. [5] using a magnetic resonance technique, corroborated da Vinci's findings when he reported that healthy human hearts eject a spiral flow that originates in the ventricle and passes into the sinuses of Valsalva, where a portion then separates from the main ejection and forms a vortical flow with momentum that promotes the closure of the aortic leaflets (See Fig. 22.5a, b). Kilner's report contained the first MRI description of normal blood flow in the aortic arch of healthy individuals.

A couple of years later, Marinelli [6], in his refutation of the pressure propulsion premise of heart function, proposed that normal blood flow involved a transfer of momenta by vortical and helical motions, including even the spinning of individual blood cells on their axes within curvilinear streams, the shape of red cells being a testament to the concept.

"The Thrill Is Gone!"

The very first evidence of spiraling or helical flow through the On-X valve came in 2004 when Dr. Hillel Laks was digitally probing the aorta on reperfusion after implanting an On-X valve (Fig. 22.6). He exclaimed, "The thrill is gone!" describing the "thrill" as a noticeable vibration he had always felt at the instant blood first surged through a traditional mechanical prosthesis and which he had been interpreting as a disruption in the flow.

An echocardiogram produced by Aman Mahajan (a colleague of Laks at UCLA, Fig. 22.7) of the On-X aortic valve in the short-axis view revealed a split red/blue image that confirmed normal, healthy, undisturbed, spiraling blood flow. Mahajan's echocardiograms showed consistent evidence for all sizes of On-X valves. Compare

Fig. 22.6 Hillel Laks MD, Cardiothoracic Surgeon at Ronald Reagan UCLA Medical Center. (Used with the permission of Hillel Laks MD and Artivion, Inc.)

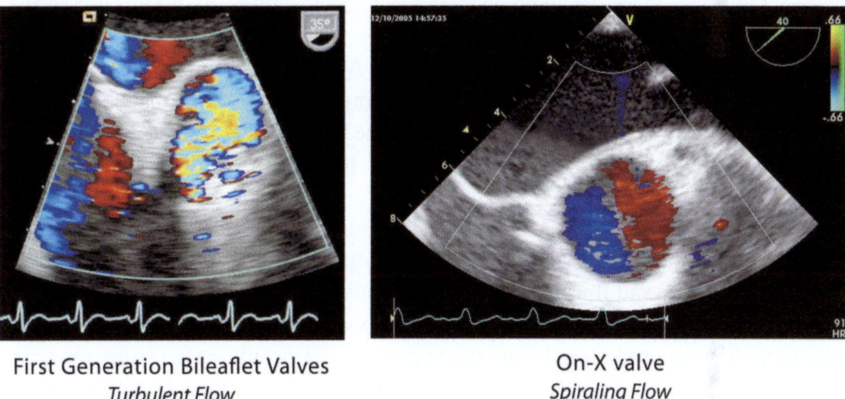

Fig. 22.7 Echocardiograms of the On-X aortic valve in the short-axis view. (Used with the permission of Hillel Laks MD and Artivion, Inc.)

this image with an echocardiogram typical of all traditional first-generation bileaflet valves, where multiple colors indicate flow disruption and abnormality.

Stonebridge [7] had already confirmed that normal blood flow is spiraling (Fig. 22.8), even in 6 mm vessels below the knee, and demonstrated its remarkable stability.

While positioned in the plane of the probe and looking upstream, the spiraling flow toward you appears red. Looking to the right downstream of the flow, the flow is moving away from you and appears blue. The red/blue image is caused by a Doppler shift that is visible.

An analogy of the Doppler shift for sound can be demonstrated by a passing train sounding its horn. As the train approaches, the pitch of the horn is high. As the train

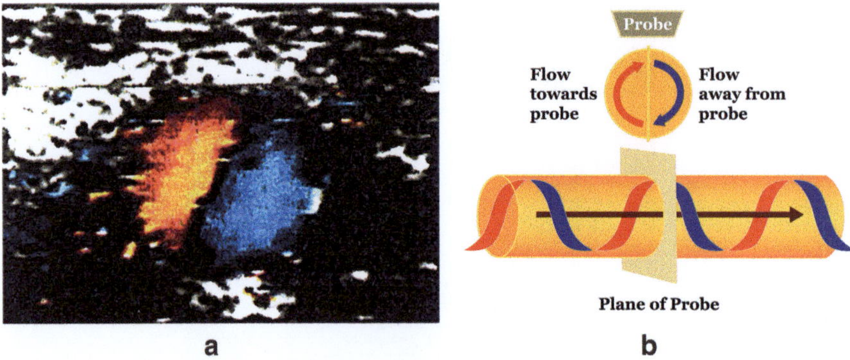

Fig. 22.8 (**a**) MRI of spiral flow in healthy 6 mm vessels below the knee. (**b**) Illustration of spiral flow. (Reprinted from *Clinical Science*, Spiral laminar flow in vivo, PA Stonebridge, et al., 1996;91(1) with permission from Portland Press, Ltd)

passes, the pitch decreases. The decrease in pitch is caused by a Doppler shift for sound that you can hear.

In the year 2000, Kilner et al., in *Letters to Nature* [8], published this extraordinary visual depiction (Fig. 22.9) achieved using a magnetic resonance technique. The imagery shows the normal flow through the healthy heart of a 34-year-old man, a flow that could be degraded or interrupted by, e.g., a heart valve replacement or repair.

Starting sequentially (Fig. 22.9) in early ventricular systole, (a) blood enters the right atrium from the superior and inferior caval veins and (b), in early ventricular diastole, continues out through the tricuspid valve to the lungs.

During ventricular systole (c), returning blood passes into the left atrium, where the streamlines are redirected from the upper and lower pulmonary veins toward the mitral valve, which is being pulled open by the twisting motion of the left ventricle, with the streamlines then being redirected (d) through the open mitral valve into the left ventricle as streamlines are ejected (e) from the left ventricle through the aortic valve. In early diastole (f), streamlines continue to pass from the left atrium through the open mitral valve.

This sequence is consistent with a preemptive opening of the mitral valve before the aortic valve closes, as described by Buckberg and his colleagues.

Kilner's group proposed that the asymmetries and curvature of the looped heart deliver potential fluidic and dynamic advantages. As a result, the inflowing momentum of the streamlines is redirected toward the valves. This momentum redirection means that the recoil away from ejected blood enhances atrio-ventricular coupling while minimizing energy dissipation.

If you recall your experiences at a carnival while riding on the tilt-a-whirl or the roller coaster, you can imagine the kind of forces of acceleration, spinning, and looping that cells must survive. These dynamic factors become especially significant when undertaking the design of an artificial replacement for a natural valve.

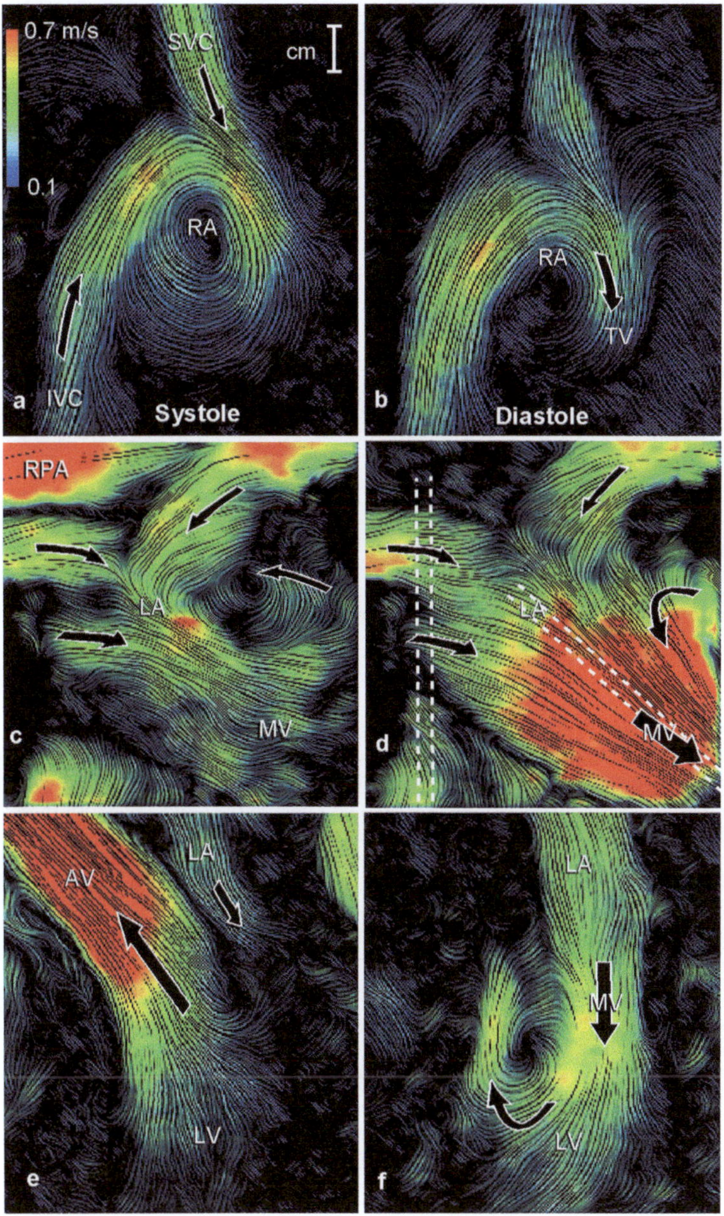

Fig. 22.9 Asymmetric redirection of flow through the healthy heart of a 34-year-old man. (Reprinted from *Nature*. Asymmetric redirection of flow through the heart, PJ Kilner, et al., 2000 with permission from Springer Nature)

Kilner warned not to be deceived by apparently smooth streamlines and pointed out the limitations of current MRI technology. MR images, he said, are created by repeated detections. As a consequence, only flow patterns that repeat sequentially from beat to beat are detectable. This means that small spinning entities that occur in a split second, below the temporal threshold of the detection rate, are completely invisible to the MRI instrument.

Thus, Marinelli's imagined spinning of individual blood cells cannot be captured by MRI. The same proviso applies to small vortical entities that arise as large vortices shed energy by dissipating into smaller ones that are invisible to MRI. These smaller, short-term vortices are ejected in the sequential contractions of the helical ventricular myocardial band (HVMB) to produce a spiral ejection that originates at the apex of the ventricle and continues out through the aortic valve.

The only valve replacement that duplicates this complex dynamic achievement of the natural valve is the On-X aortic valve, as observed in 2004 by Laks.

References

1. Zilla P, Brink J, Human P, Bezuidenhout D. Prosthetic heart valves: catering for the few. Biomaterials. 2008;29(4):385–406.
2. Summary of safety and effectiveness for On-X 2001 aortic, FDA PMA P000037, May 30, 2001 and March 6, 2002 European Primary Trial Updated May 31, 2003.
3. Leonardo da Vinci, An Artabras Book. Reynal and Company in association with William Morrow and Company, New York. All rights reserved under International and Pan-American Copyright Conventions and copyright by Instituto Geografico De Agostini, Novara, Italy.
4. O'Malley CD, Saunders JBDCM. Leonardo da Vinci on the human body. New York: Wings Books. Avenel Random House ISBN 0-517-38105-2.
5. Kilner PJ, Yang GZ, Mohiaddin RH, et al. Helical and retrograde secondary flow patterns in the aortic arch studied by three-directional magnetic resonance velocity mapping. Circulation. 1993;88:2235–47.
6. Marinelli R, Fuerst B, Vander H, et al. The Heart is not a pump: a refutation of the pressure propulsion premise of heart function. Front Persp. 1995;5(1).
7. Stonebridge PA, Buckley C, Thompson D, et al. Non-spiral and spiral flow patterns; in-vitro observations using magnetic resonance imaging and computational fluid dynamic modeling. Int Angiol. 2004;23(3):276–83.
8. Kilner PJ, Yang GZ, Wilkes AJ, Mohladdin RH, Firmin DN, Yacoub MH. Nature. 2000;404(6779):759–61.

Chapter 23
Studies Concurrent with FDA Trials

Clinical trials carried out in Europe and the United States demonstrated low complications and exceptional hemodynamics. In 2001, these results led to FDA approvals of the On-X aortic valve and in 2002 of the On-X mitral valve [1, 2]. These achievements were rightly attributed to John Ely.

While our clinical FDA trials were progressing in Europe, the Early Self-Controlled Anticoagulation Trial (ESCAT) [3] of anticoagulation therapy by Körtke and Körfer was underway in nine centers in Germany. As the On-X valve did not yet have FDA approval, it was not allowed to be a candidate in the study. To further expand our non-FDA trials, Ely elected to follow the same ESCAT protocol in German patients receiving On-X valves. The valves used in the ESCAT were St. Jude Medical; CarboMedics, Inc.; and Medtronic-Hall.

When the first 600 patients from the ESCAT completed a 2-year follow-up, the ESCAT had 295 conventionally monitored patients. The conventionally monitored patients with On-X valves numbered 322. The patient demographics (Fig. 23.1) were similar in all respects, except the On-X mitrals were proportionally higher (36% vs 19%). The number of patients with atrial fibrillation was 27.5% for On-X and 25.5% for ESCAT. Only Grade III complications were compared across the sample base.

Grade III patients included those with prosthetic thrombosis and severe thromboembolism, debilitating conditions requiring inpatient treatment and usually inflicting long-term impediments (including transient ischemic attacks). This group also included cases of severe hemorrhage requiring transfusion, surgical or endoscopic intervention, or in-patient care, with long-term impediments.

Patients with On-X valves experienced one-third of the Grade III complications that occurred in the ESCAT (Figs. 23.2 and 23.3). The INR distributions are tabulated in Fig. 23.4.

Details of the complicated ESCAT were published in Medical Carbon Research Institute's newsletter, *The On-X Experience*. The final draft by Kathleen Selbrede was transmitted to Endre Bodnar for his review. Endre's response was: "This draft is suitable for publication as it stands."

J. Bokros, *Heart of Carbon*, https://doi.org/10.1007/978-3-031-17933-4_23

	ON-X	**ESCAT**
Average Age (yr.)	60.1 10.6	62.5 10.2
Female %	43	34
Male %	57	66
Mitral %	36	18.5
Aortic %	57	74.5
Double %	7	7
Atrial Fibrillation %	27.5	25.5

Fig. 23.1 ESCAT patient demographics. (Used with the permission of Artivion, Inc.)

	ON-X CONVENTIONAL	**ESCAT** CONVENTIONAL	**ESCAT** SELF-MANAGED
AVG FOLLOW-UP (yrs)	2.8	3.2	3.2
FOLLOW-UP (pt-yrs)	885	943	976
HEMORRHAGIC N (% per pt-yr)	5 (0.6)	25 (2.6)	17 (1.7)
THROMBOEMBOLIC N (% per pt-yr)	8 (0.9)	20 (2.1)	12 (1.2)
TOTAL N (% per pt-yr)	13 (1.5)	45 (4.7)	29 (2.9)

Fig. 23.2 ESCAT Grade III complications. (Used with the permission of Artivion, Inc.)

Mervyn Williams and the South African Study

At the same time, studies of On-X carbon valves in a non-compliant South African population were initiated in 1999. Mervyn Williams, MD, of Port Elisabeth in South Africa, presided over the accumulation of this highly significant data. Kinsley, a colleague of Williams, pointed out how conducting such trials in developing nations provides an excellent baseline for determining the reliability and thrombogenicity of new valve designs [4].

Kinsley's idea is sound so far as it goes, but translating it from concept to reality requires an infrastructure be in place to permit following the data to properly grounded conclusions.

In South Africa, those key prerequisites actually *were* in place. Mervyn's wife Bernadette, an RN and Mervyn's nurse, followed the patients. This was not an easy task because so much of the communication was "conducted via Land Rover."

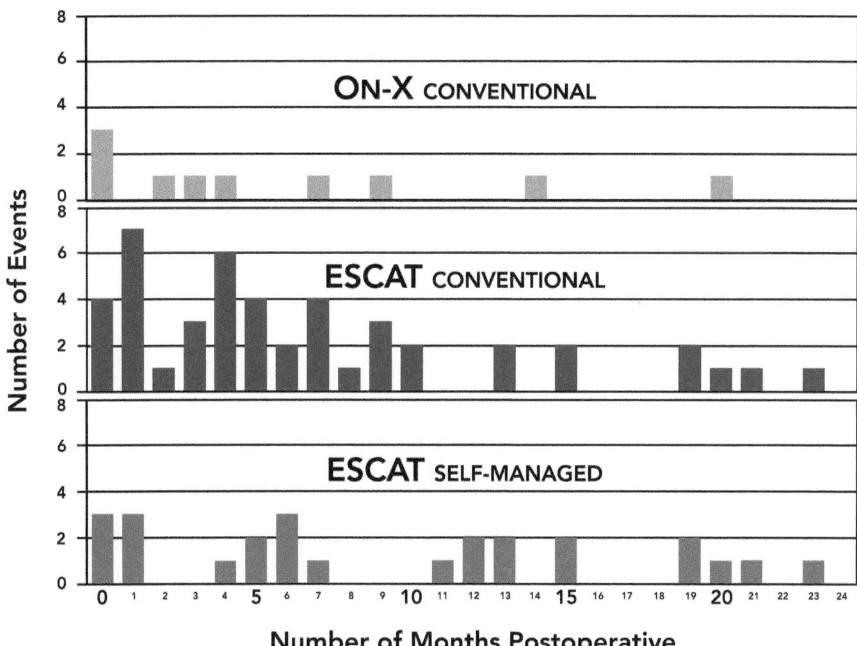

Fig. 23.3 ESCAT Grade III complications for the first 24 months post-operatively. (Used with the permission of Artivion, Inc.)

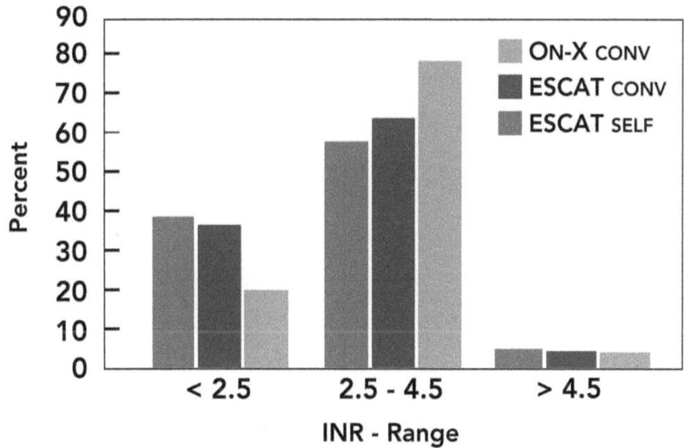

Fig. 23.4 ESCAT INR distributions for the On-X/ESCAT (St. Jude Medical; CarboMedics, Inc.; and Medtronic-Hall valves) comparison. (Used with the permission of Artivion, Inc.)

Bernadette came up to me at a conference and said, "Dr. Bokros, I want you to assure me that the On-X valve will be better than the CarboMedics, Inc. valve."

"Yes, indeed," I answered. "It will be much better."

She had entreated me because Mervyn had previously used the CarboMedics, Inc. valve and had been unhappy with the results.

Between October, 1999, and November, 2014, Mervyn Williams implanted 1387 valves in a South African patient population, many of whom did not follow the designated anticoagulation protocol. Forty-seven percent were anticoagulated within an INR of 1.5–2.5. Fifty-three percent were not anticoagulated, or their anticoagulation was outside the planned range.

In January of 2017, Williams informed me that his analysis of the data was complete, and he provided me a confidential copy of a paper he had submitted for journal review. In 2014, the On-X valve 15-year trial in a poorly anticoagulated population was closed. The data were presented at the annual meeting of the Asian Society for Cardiovascular and Thoracic Surgery in Seoul, South Korea, in March 2017 but never published.

Belgium Reports In

In Belgium, Flameng (Fig. 23.5) [5], a highly respected investigator, served as the gatekeeper for valves coming into Belgium. He carried out his valve studies in sheep. The thromboresistance of valves was determined by implanting six valves in the pulmonary position, without anticoagulation, and monitoring the sheep for

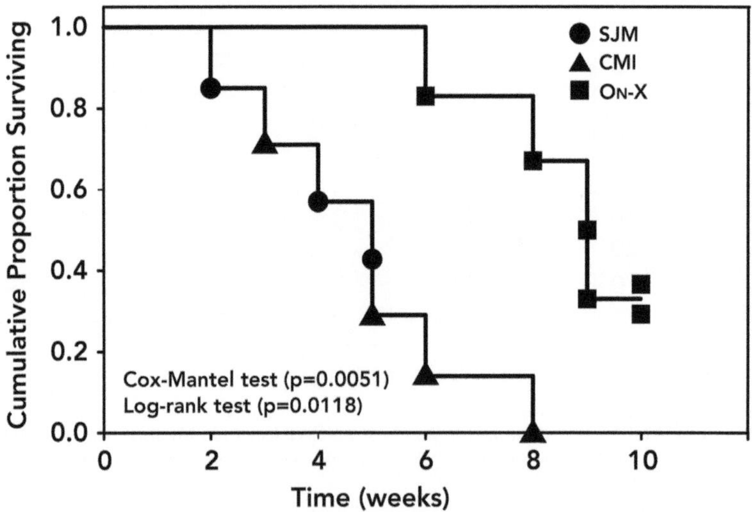

Fig. 23.5 Results for six On-X valves implanted in the pulmonary position in sheep without anticoagulation monitored for 10 weeks maximum for thrombotic complications and compared with results for three St. Jude Medical and four CarboMedics, Inc. valves. (Used with the permission of Artivion, Inc.)

thrombotic complications for 10 weeks. In 2002, he presented the data comparing CarboMedics, St. Jude Medical, and On-X valves and found that the thromboresistance of On-X valves was not only high; it measured better than the St. Jude Medical and CarboMedics valves ($p<0.01$).

Putting the Pieces All Together

In 2000, with data from the US and European clinical trials, the data from South Africa, the ESCAT study, and the Flameng sheep data, Ely approached the FDA and proposed an aspirin-only study. The FDA recommended that an aspirin-only feasibility study be carried out first.

Accordingly, Ely, alongside Professor Uwe Mehlhorn of the University of Cologne Heart Center, approached the German Regulatory Authority and obtained approval for use of 100 mg aspirin at six centers in Germany using aortic patients whose only risk factor was the valve. The study was to last for 5 years.

In 2005, when the study was in midstream, Ely again approached the FDA and obtained an IDE approval to carry out a reduced anticoagulation intensity study that included all patients, e.g., not just aortic ones.

This study, to be conducted at Emory University with John Puskas, MD, as Principal Investigator, had three arms: (1) low-risk aortic patients with the valve being the only risk factor, to be anticoagulated only with platelet inhibitors; (2) high-risk aortic patients to receive 81 mg aspirin and Coumadin with INR = 1.5 to 2.0; and (3) mitral patients to receive an 81 mg aspirin with Coumadin at INR = 2.0 to 2.5 (Fig. 23.6a, b).

The trial, called the PROACT, was initiated in 2006. The number of Medical Carbon Research Institute personnel at the time (Figure 23.7) can be compared to the team employed at Medical Carbon Research Institute's birth in 1994 (Fig. 20.2 in Chap. 20).

In 2005, using an analysis originated by Bodnar, we focused on thrombotic and bleeding complication rates and their dependence on INR. Clinical data for mechanical valves are depicted in Fig. 23.8a, b. The solid black line in Fig. 23.8a represents bleeding. The upper dark gray line represents mechanical valves used prior to the 1980s without carbon [6]. The broad lighter gray band includes clinical data for Pyrolite Carbon co-deposited with silicon. The very light gray band is an estimation using the On-X valve in a non-compliant population in South Africa coming from Mervyn Williams, MD, in Port Elisabeth, South Africa.

A 1998 review by Butchart [7] entitled "Prosthetic Heart Valves" is reproduced here as Fig. 23.8b. Figure 23.8b includes a curve labeled *Valve C* for the Medtronic, Inc. Parallel Valve. *Valve C* is presented as being susceptible to thrombose. Curves labeled A and B contain less thrombogenic valves with A the most thromboresistant. Data coming from Medtronic, Inc. was never reported, so the clinical use of the Parallel Valve in humans was never disclosed.

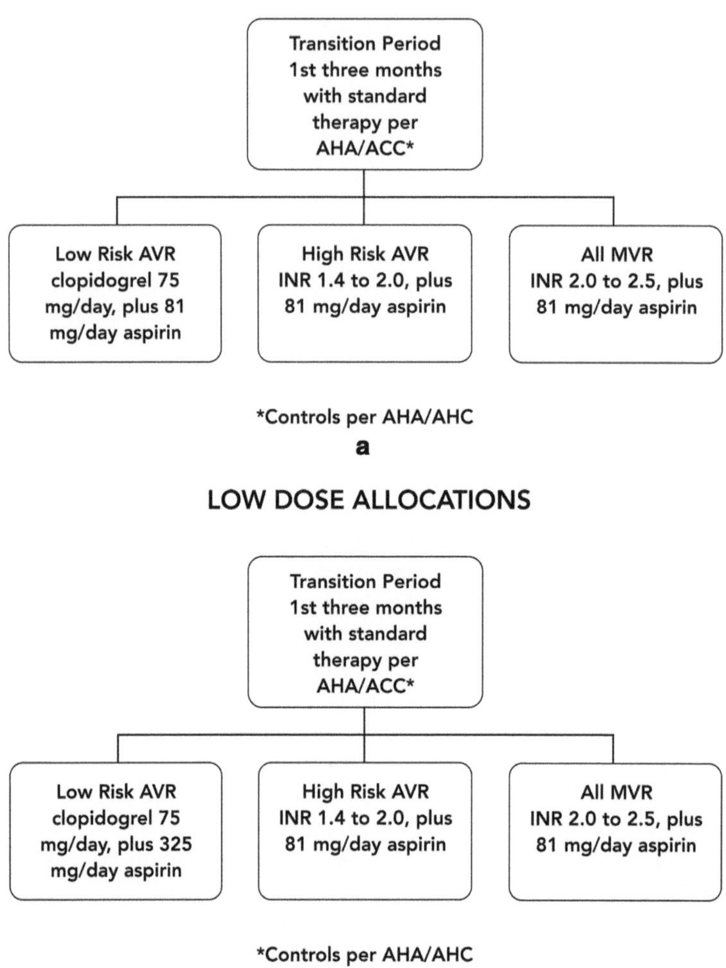

Fig. 23.6 (**a**) Prospective Research for the On-X Anticoagulation Trial (PROACT) initial patient allocations done after a 3-month transition period using standard therapy per AHA/ACC controls. (**b**) Aspirin dose for low-risk aortic valve replacement (AVR) was increased from 81 to 325 mg/day with FDA concurrence because of high rates of patient non-response to aspirin. (Used with the permission of Artivion, Inc.)

Lacking a report from Medtronic, Inc. on the Parallel Valve, the profession was only aware of the pre-clinical testing, the results of which were unequivocally outstanding, ranking the valve on a scale of 1 to 10 as an 11. The reports of clotting, circulating as hearsay, asserted that the valve must actually be thrombotic. But unless the Medtronic, Inc. Parallel Trial had followed the guidelines set forth by the

Fig. 23.7 Personnel of Medical Carbon Research Institute when the On-X PROACT was initiated. (Used with the permission of Jack Bokros)

American College of Chest Physicians and the National Heart, Lung, and Blood Institute, then and only then could it be legitimately concluded that the valve was susceptible to clotting.

The profession, lacking input describing the detailed clinical results, assumed that the Trial *did* follow the guidelines set forth by the American College of Chest Physicians and NIH. The perception was that the guidelines were appropriate, but the protocols actually used were simply not disclosed. The facts beg for an answer about the *actual* thrombogenicity of the Parallel Valve when tested on a level playing field.

Promoting the On-X Valve

Rollie Siegel and I joined with distributors in the United States and internationally (through Regina Burnett, VP of International Sales) to promote On-X valve sales. Rollie had already introduced us to Dr. Sidney Levitsky (Fig. 23.9), a senior Professor at the Brigham and Women's Hospital in Boston. John Ely appointed him to an oversight position, Chairman of the Data Safety Monitoring Board. Levitsky was to periodically review the clinical data generated in the PROACT and certify that it was safe to proceed. He even traveled to South Africa to collect EKG data on On-X valve patients.

The venture capitalists were impressed by Ely's having gained an IDE for a low-dose anticoagulation trial. As a first ever, the venture capitalists and their executives

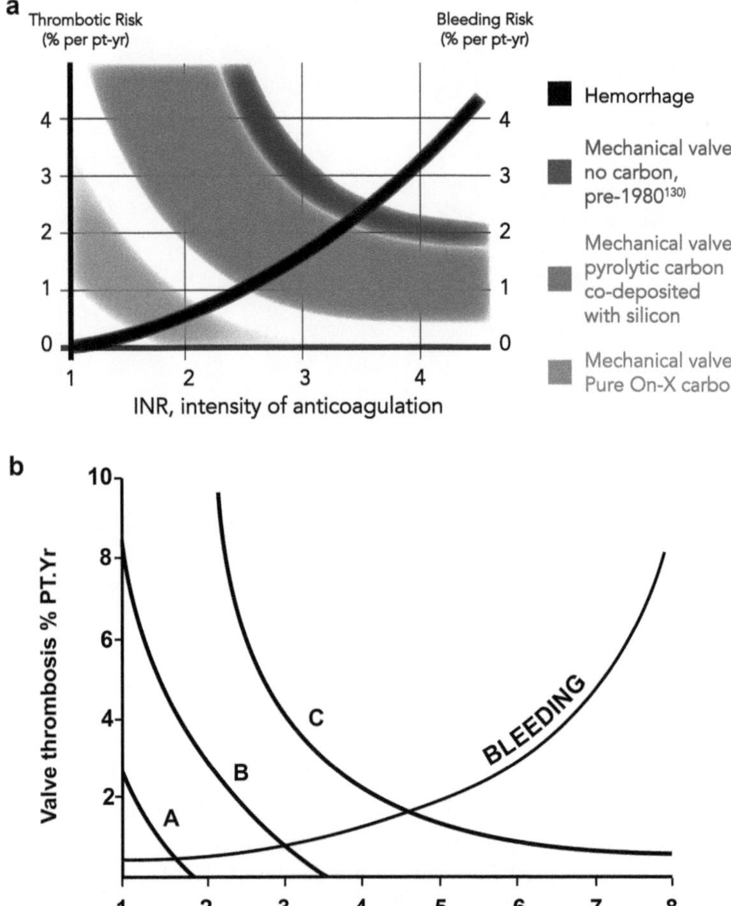

Fig. 23.8 (**a**) Schematic forecast for On-X valves based on published data along with data being collected by Mervyn Williams MD in the On-X South African Trial. The actual position of the On-X curve was later established after the completion of the On-X PROACT. (Used with the permission of Artivion, Inc.). (**b**) Thrombogenicity curves of three hypothetical valves relating valve thrombosis rate to anticoagulation intensity. See text for the detailed explanation of labeling error; Valve C (Parallel Valve) is labeled as hyperthrombogenic but its thromboresistance was unknown at that time. (Courtesy of Eric G Butchart)

predicted that the announcement would accelerate sales. Although sales were indeed increasing, the rate of increase was low.

In 2004, 2005, and 2006, Medical Carbon Research Institute's valve sales were worth about 9 million, 12 million, and 15 million dollars, respectively. The growth in sales was driven by the FDA approvals in 2001 and 2002, together with the non-FDA studies from the ESCAT, Flameng's sheep data, results achieved by Williams in South Africa for a non-compliant population, and the aspirin-only study in Germany for healthy aortic patients with the valve as the only risk factor.

Fig. 23.9 Dr. Sidney Levitsky at Brigham and Women's Hospital in Boston, MA, was the Chairman of the Safety Monitoring Committee for the PROACT. (Used with the permission of Dr. Sidney Levitsky)

Fig. 23.10 G Russell Chambers, President and CEO of Intermedics Inc. with Steven G Anderson, Senior Executive Vice President on the right. (Image provided courtesy of Boston Scientific ©2019 Boston Scientific Corporation or its affiliates. All rights reserved. Also courtesy of Steven Anderson)

Even though the On-X valve had already provided evidence that it stood alone as the only valve replacement to preserve normal flow through the valve, its sales were lagging because I was convinced there were deficiencies in marketing.

In the 1980s, sales by Intermedics, Inc. Pacers (led by Executive Vice President Steve Anderson, Sr., Fig. 23.10) had been tripling annually. In fact, in 2002, when we had FDA approval under our belt but felt an inadequacy in marketing, I turned to Anderson, then President and CEO of CryoLife. On invitation, I traveled to Kennesaw, Georgia, to entice Anderson to consider marketing Medical Carbon Research Institute's On-X valves.

CryoLife's facility reflected Anderson himself, with a video studio and a warehouse filled with "marketing machines." The meeting was reciprocated by a visit to Medical Carbon Research Institute by Anderson and a financial executive to evaluate the prospects.

Steve was genuinely interested. After a dialogue, I laid a 40% margin on the table. He smiled and declared "you've got that backwards." I was set back on my heels by his reply but remained silent. Steve advised me that they'd think about it and get back to us.

While awaiting his response, I considered coming back to the table with a 50% margin. Considering Steve's market paraphernalia, 60% might actually have worked. But the bottom line was that Steve and his team could not rationalize a fit with Medical Carbon Research Institute. However, that was by no means the end of the story with CryoLife.

Siegel and I spent our time promoting the On-X valve to surgeons, emphasizing first the unique hemodynamic design and then using the lead offered by Hillel Laks's echo evidence that the On-X valve allowed the normal helical ejection through the aortic valve to proceed into the sinuses of Valsalva without disruption.

Our template for presentations focused on the unique design, which was technologically inaccessible by other valve replacements, and then demonstrating what constitutes the normal flow that the On-X valve is actually able to preserve.

Leveraging that premise, we could affirm that "we're the only valve ever granted an IDE for a low dose study." The surgeons would naturally reply, "If you have a valve that preserves normal flow, that would be entirely expected." But without that key premise, their deep skepticism would kick in: "That's nice, but come back once you have the approvals." Without the preservation of normal flow, you're standing on a marshy ground.

References

1. Summary of safety and effectiveness for On-X 2001 aortic, FDA PMA P000037, May 30, 2001 and March 6, 2002 European Primary Trial Updated May 31, 2003.
2. Summary of safety and effectiveness for On-X 2002 mitral, FDA PMA P000037/S1, May 30, 2001 and March 6, 2002 European Primary Trial Updated May 31, 2003.
3. Körtke H, Körfer R. International normalized ratio self-management after mechanical heart valve replacement: is an early start advantageous? Ann Thorac Surg. 2001;72:44–8.
4. Kinsley RH, Colsen PR, Antunes MJ. Medtronic-Hall replacement in a third-world population group. Thorac Cardiovasc Surg. 1983;31(11):69.

5. Meuris B, Flameng W. Performance of bileaflet heart valve prostheses in a new animal model. In: Presented at the 6th Annual Hilton Head Workshop of prosthetic heart valve. March 6–10, 2002.
6. Cannegieter SC, Rosendall FR, Briet B. Thomboembolic and bleeding complications in patients with mechanical heart valve prostheses. Circulation. 1994;89:635–64.
7. Butchart EG. Chapter 22: Prosthetic heart valves. In: Verstracete M, Fuster V, Topol EJ, editors. Cardiovascular thrombosis: thrombocardiology and thromboneurology. 2nd ed. Philadelphia: Lippincott-Raven Publishers; 1998. p. 395–414.

Chapter 24
Understanding Cardiac Blood Flow

Starting with Kilner's 1993 [1] MRI defining normal blood flow and extending to more detailed imaging in 2000 [2], rapid advances in identifying the origin of blood in the left ventricle have been reported by Gerald Buckberg. An important compilation appeared in 2006 as Supplement 1, Volume 9, 2006, of the *European Journal of Cardio-Thoracic Surgery* featuring Buckberg as guest editor. In that Supplement, a paper by Bockeria provides a connection between Laks's echo observations and the physiological studies of da Vinci [3].

The paper contains a page from Leonardo da Vinci's manuscripts, showing spiral patterns that matched the morphology of the trabeculae on the left ventricular wall. A photograph from the Bockeria paper is included as Fig. 24.1.

A similar observation from da Vinci was displayed later by Ares Pasipoularides on page 122 of his book *Heart's Vortex*. That seminal reference depicts flow patterns within the sinuses and describes a glass model that da Vinci used to study blood flow in the sinuses [4].

Another Supplement 1 paper (Coghlan) [5] describes an important advance made in 1957 by Torrent-Guasp, who performed a simple but critical experiment. In a blunt (hands-only) dissection of the heart, he demonstrated that the heart could be unfolded, revealing it to be a long tube, with the aorta on one end and the pulmonary artery on the other, Fig. 24.2 (from Ref. [5]). The right end displays the descending and ascending segments of the HVMB. The HVMB is a complex musculature that is responsible for producing a helical ventricular ejection out of the heart through the aortic valve.

Gerald Buckberg and the Advance of Heart Science

After 1957, it took 60 years to take the next step in understanding the form and function of the heart as laid out in Buckberg's Supplement. The function of the HVMB is shown schematically in Fig. 24.3 [5].

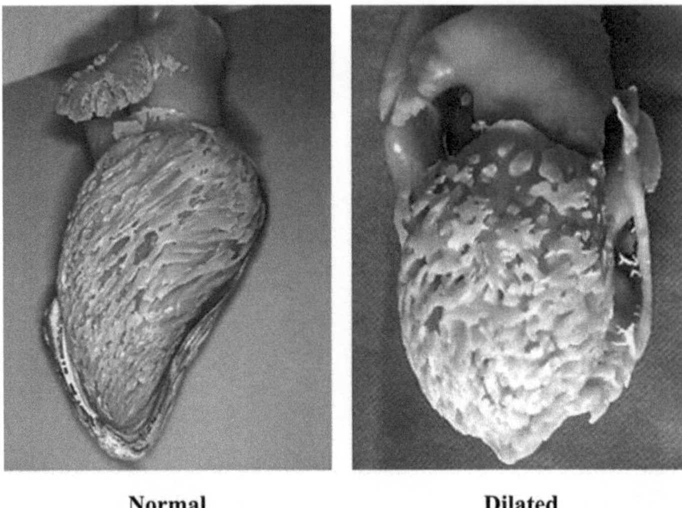

Fig. 24.1 Plaster casts of the left ventricle: normal and dilated. Note the oblique spiral trabeculae in the normal heart and the development of a more horizontal pattern in the dilated heart before ventricular restoration. (Reprinted from the *European Journal of Cardio-Thoracic Surgery*. Left ventricular geometry reconstruction in ischemic cardiomyopathy patients with predominantly hypokinetic left ventricle, LA Bockeria, AJ Gorodkov, 2006;29(S1) with the permission of Oxford University Press)

Gerald Buckberg provided a simplified schematic representation of the complex sequential ventricular rotations that produce spiral ejection of blood from the ventricle. During systole, the ascending segment (sloping upward to the left and overlapping the descending segment) of the HVMB starts to contract; then the descending segment starts to contract. As the latter segment continues to contract to maximum excursion, the ascending segment continues contracting, so that the heart's ejection takes a helical form (Fig. 24.4) [6]. Final contraction of the ascending segment, called a hiatus, creates a decrease in pressure in the ventricle, which is thought to assist ventricular filling.

The mapping of spiral flow published by Kilner depicted in Fig. 22.5a, b [1] shows that the spiral ejection persists through systole to the closure of the aortic leaflets, thus preserving turbulent-free, orderly spiral flow. At the end of systole, the persisting vortical flow in the sinuses propels the flow into the coronaries.

In 2004, Stonebridge et al., as mentioned earlier, added to the understanding of spiral flow. He and his colleagues cited a technical note that is important because it provides data comparing the relative stability of spiral and straight streamlined flows [7].

The technical note compared helical and non-spiral streamline flows in 10 mm glass tubes. In smooth tubes, the flows were the same. But when an obstacle was placed in the tube, the spiral flow maneuvered around the obstruction and continued on with little energy loss.

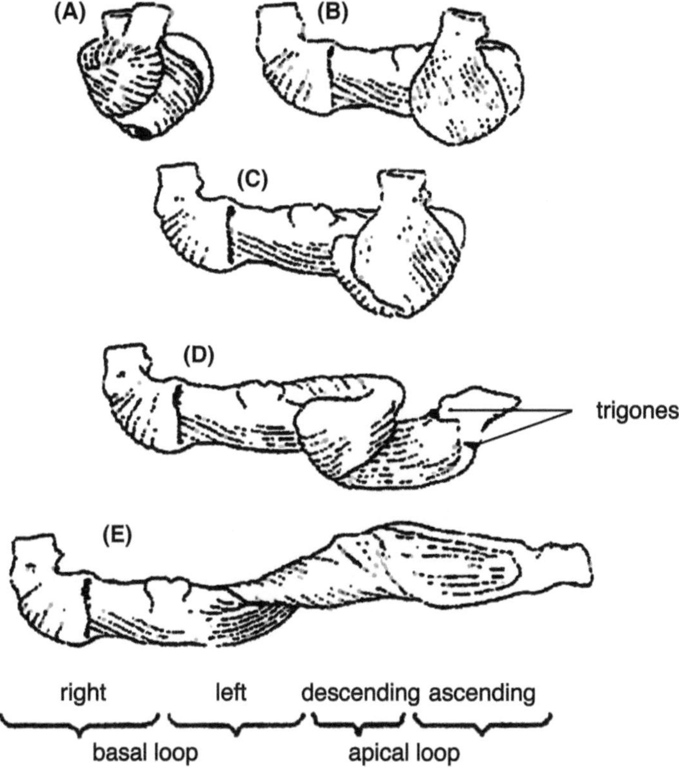

Fig. 24.2 Unscrolling of the HVMB. (Reprinted from the *European Journal of Cardio-Thoracic Surgery*. Leonardo da Vinci's flights of the mind must continue: cardiac architecture and the fundamental relation of form and function revisited, C Coghlan, J Hoffman, 2006;29(S1) with the permission of Oxford University Press)

In the case of non-spiral streamline flow, when the flow encountered an obstacle, it lost continuity and produced turbulence that was evaluated using magnetic resonance imaging and computational fluid dynamics modeling. The Stonebridge team calculated that in the case of non-spiral streamlined flow, the encounter with an obstacle caused the release of *seven times more turbulent energy near the wall* than was released when spiral flow encountered the same obstacle. This served as a clear demonstration that natural spiraling flow gains a stability advantage over streamlined non-spiraling flow.

Following this lead, Stonebridge developed a graft with an internal spiraling rib. This construct introduced spiral flow to increase flow stability, making the distal anastomosis (downstream connection of one vessel to another) more thromboresistant.

A paper by Bakhtiary (2006) and his colleagues [8] is equally relevant. The investigators measured coronary flow reserve after aortic valve replacement. The study embraced four valves: the Medtronic-Hall, Medtronic Advantage, Medtronic

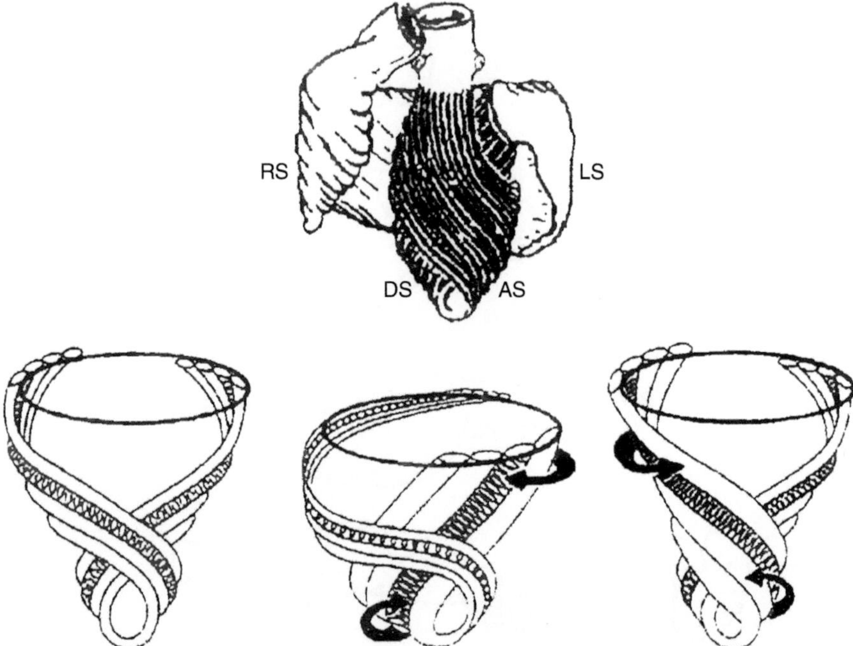

C. Coghlan et al. European Journal Cardio-thoracic Surgery 2006;295:S4-S17

Fig. 24.3 Schematic representation of the complex sequential ventricular rotation that results in spiral flow. The sequential contraction of the ascending segment begins before the contraction of the descending segment. The ascending segment continues contracting after the descending segment stops contracting, which reduces the ventricular pressure and results in a spiral ejection. (Reprinted from the *European Journal of Cardio-Thoracic Surgery*. Leonardo da Vinci's flights of the mind must continue: cardiac architecture and the fundamental relation of form and function revisited, C Coghlan, J Hoffman, 2006;29(S1) with the permission of Oxford University Press)

Mosaic, and the Medtronic FreeStyle. There were 4 groups of 12 patients, each with aortic stenosis (abnormal narrowing of the aortic valve), who received 1 of the 4 prostheses listed above. Normalization of coronary flow reserve was only observed for the stentless valve.

The authors attributed the good results for the Medtronic FreeStyle stentless bio-prosthesis to the natural shape of stentless prostheses, which provided non-turbulent normal flow into the coronaries.

Data from FDA submissions in Fig. 24.5 [9] compare the effective orifice area of On-X valves, size for size, to corresponding data for the FreeStyle. There is better flow, size for size, for the On-X valves. The implication is that the On-X valve provides better flow into the coronaries, but this fact has yet to be empirically demonstrated.

Coronary flow reserve is an important metric for predicting long-term survival after aortic valve replacement for aortic stenosis. Reduced coronary flow reserve contributes to more frequent cardiac events and greater rates of mortality.

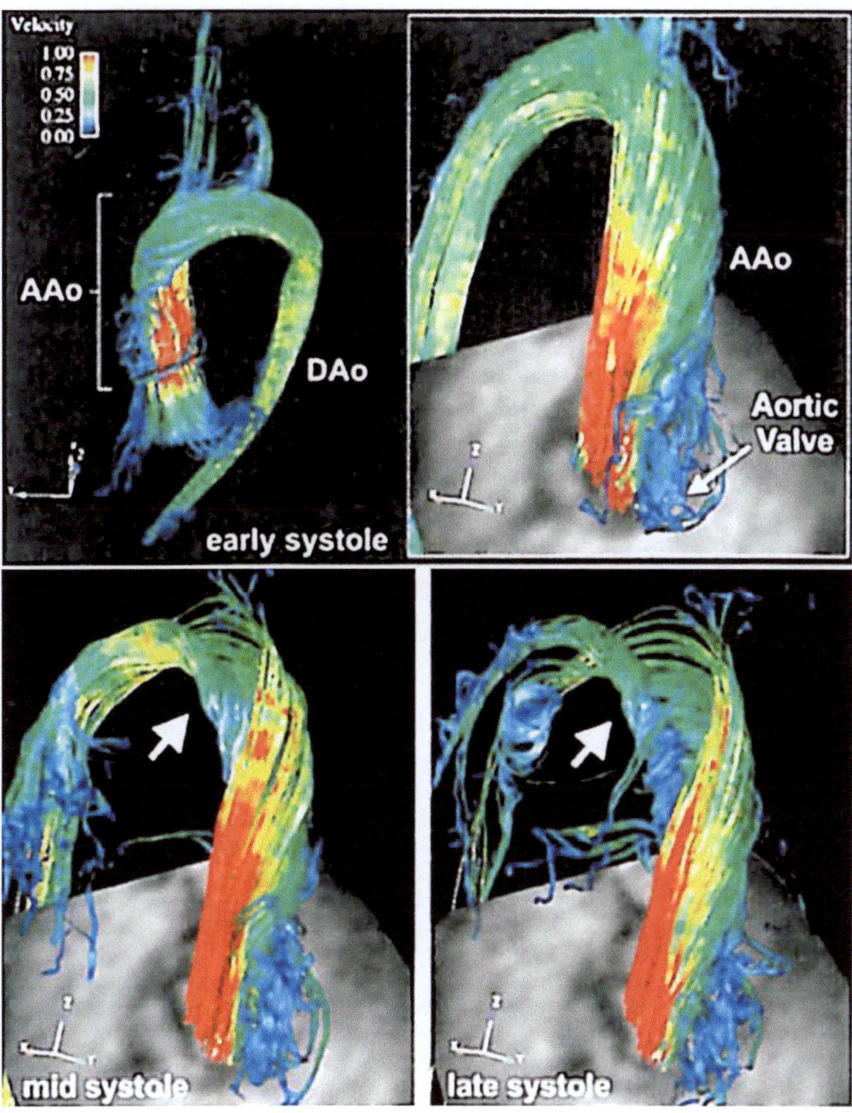

Fig. 24.4 Right-handed spiraling flow in the ascending aortic arch during mid- and late systole. (Reprinted from the *Journal of Thoracic and Cardiovascular Surgery*. Time-resolved three-dimensional magnetic resonance velocity mapping of aortic flow in healthy volunteers and patients after valve-sparing aortic root replacement, M Markl, et al., 2005;130(2) with the permission of Elsevier)

The rapid development of our understanding of heart function and normal flow, as described in Buckberg's Supplement, solidified our presentation. As a result, the key premise—the unique On-X design and the valve's ability to preserve normal flow—became compelling enough in tandem to overcome skepticism and resistance by providing surgeons the assurance that reduced anticoagulation is an obvious expectation.

Fig. 24.5 Comparison of
FDA submission data.
(Used with the permission
of Artivion, Inc.)

FDA SUBMISSION DATA

Efffective Orifice Areas for Aortic Valves

Valve Size	On-X	Freestyle
19	1.5	1.0
21	1.8	1.4
23	2.3	2.0
25	2.7	2.2
27	2.9	2.4

References

1. Kilner PJ, Yang GZ, Mohiaddin RH, et al. Helical and retrograde secondary flow patterns in the aortic arch studied by three-directional magnetic resonance velocity mapping. Circulation. 1993;88:2235–47.
2. Kilner PJ, Yang GZ, Wilkes AJ, Mohladdin RH, Firmin DN, Yacoub MH. Nature. 2000;404(6779):759–61.
3. Bockeria LA, Gorodkov AJ, Dorofeev AV, et al. Left ventricular geometry reconstruction in ischemic cardiomyopathy patients with predominantly hypokinetic left ventricle. Eur J Cardiothorac Surg Suppl. 2006:S251–8.
4. Pasipoularides A. Heart's vortex. Shelton: Peoples Medical Publishing House; 2010.
5. Coghlan C, Hoffman J. Eur J Cardiothoracic Surg Supplement. 2006;295:S4–S17.
6. Markl M, Draney MT, Miller DC, et al. Time-resolved three-dimensional magnetic resonance velocity mapping of aortic flow in healthy volunteers and patients after valve-sparing aortic root replacement. J Thorac Cardiovasc Surg. 2005;130:456–63.
7. Stonebridge PA, Buckley C, Thompson D, et al. Non-spiral and spiral flow patterns; in-vitro observations using magnetic resonance imaging and computational fluid dynamic modeling. Int Angiol. 2004;23(3):276–83.
8. Bakhtiary F, Schiemann M, Dzemali O, et al. Impact of patient-prosthesis mismatch and aortic valve design on coronary flow reserve after aortic valve replacement. J Am College Cardiology. 2007;49(7):790–6.
9. Burnett C. Comparisons of valve performance, FDA submission data Medtronic Freestyle® Aortic Root Prostheses Summary of safety and effectiveness data submitted to the United States Food and Drug Administration. PMA P97 0031. Approval date November 26, 1997. On-X Life Technologies, Inc. (technical report).

Chapter 25
Investigating the Viability of Bioprostheses

In 2007, David Williams (the editor of the journal *Biomaterials*) posted an online preprint of an article titled "Leading Opinion" on prosthetic heart valves [1]. The lead author, Peter Zilla, an academic with an expanding career in clinical science, presented a comprehensive review of prosthetic valves focused on those derived from "fixed" biological tissue. The author's critique, which was perhaps intentionally provocative, established an assessment of bioprostheses that revealed their shortcomings and lack of long-term viability.

After reviewing the 22-page article, I immediately ordered several hundred reprints to forward to Siegel for use as an important resource.

My enthusiasm stemmed from Zilla's academic credentials. He received an MD in 1980 from the University of Vienna and then a DMed from the University of Zurich in 1983. Continuing studies resulted in earned PhDs from the University of Vienna in 1989 and the University of Cape Town in 1990. A quick study, indeed!

Zilla held the Christiaan Barnard Chair at the University of Cape Town and was Head of Cardiothoracic Surgery at Groote Schuur Hospital. He advanced to a full Professor in 1999, the year that Port Elisabeth's Mervyn Williams started implanting the On-X valve in Medical Carbon Research Institute's South African clinical study of his non-compliant patient population. By that time, Zilla and Medical Carbon Research Institute were well acquainted with one another. Zilla's "Leading Opinion" paper was published in 2008 in the *Biomaterials* journal [1].

The Zilla paper's major focus was to study bioprostheses and evaluate their viability for use in young South African patients. His studies were comprehensive. The findings were disappointing. But the findings provide incentive for competing with bioprostheses. A pertinent summary is provided below. The references in brackets, [ ], are from Zilla's Ref. [1]:

In the 1960s, Carpentier's glutaraldehyde process for fixing biological tissue became the basis for bioprosthesis development. The process remained the basis of the industry since the 1960s.

J. Bokros, *Heart of Carbon*, https://doi.org/10.1007/978-3-031-17933-4_25

The Zilla paper asserted that industry was misdirected and became entrenched in and committed to the 1960s technology instead of recognizing the limitations of non-vital, antigenic materials.

The fundamental structural issues were that healthy natural valves can repair wear and tear, had annulus flexibility reducing stresses at the commissures [106]. The tri-layer leaflet structure reduces bending stresses. Cross-linking negates all three functions.

A case in point, "stentless" concepts overstate the biomechanical advantage. Expected durability was not achieved [84, 85]. Again cross-linking is the confounder.

Glutaraldehyde intrinsically elicits calcification (Fig. 25.1).

Early researchers believed glutaraldehyde-fixed tissue would be acceptable if the right anticalcification treatment could be found, even though calcification affected less than half of failed valves [81–83]; the cause of most of the failures was not addressed.

A flurry of marketable "anticalcification treatments" reduced but did not eliminate calcification [84, 85, 108–111].

The neglected villain was remnant inflammatory/immune processes [96].

Non-calcific failures (Fig. 25.2) *were attributed to remnant antigenicity. The majority of explanted valves involve inflammatory cells. The sequential processes include:*

1. *Polymorphonuclear infiltrates [90, 103].*
2. *A macrophage and foreign body giant cell-dominated phenomenon [90, 97, 100].*
3. *Macrophage-mediated degeneration with collagen phagocytosis [99, 103, 104].*

Eighty-two percent showed collagen phagocytosis [104].

Figure 25.3 shows on the left an original collagen fiber compared with a degenerated fiber on the right. Remnant inflammatory/immune processes [96] or an inflammatory/immune process initiated by mechanical failure was proposed.

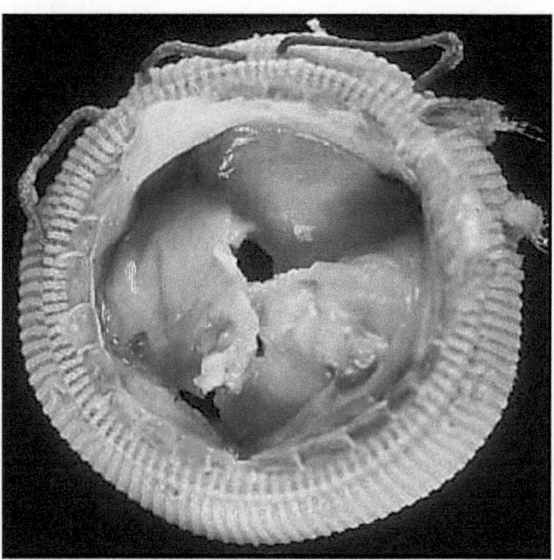

Fig. 25.1 A calcified glutaraldehyde bioprosthesis. (Reproduced with the permission of Frederick Schoen MD PhD)

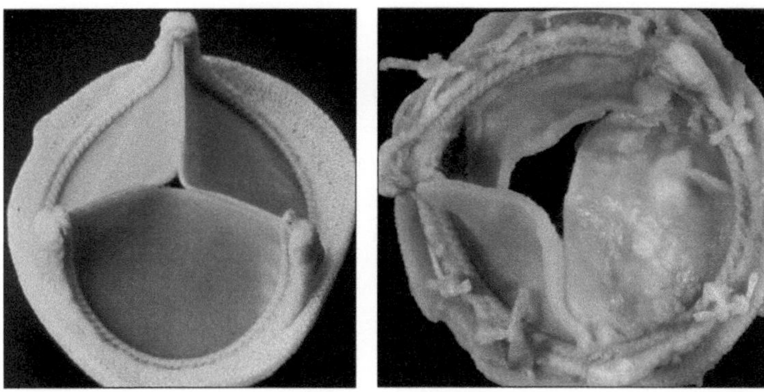

Fig. 25.2 A bovine pericardial bioprosthesis before and after implantation; an example of a non-calcific failure on the right. (Reproduced with the permission of Frederick Schoen MD PhD)

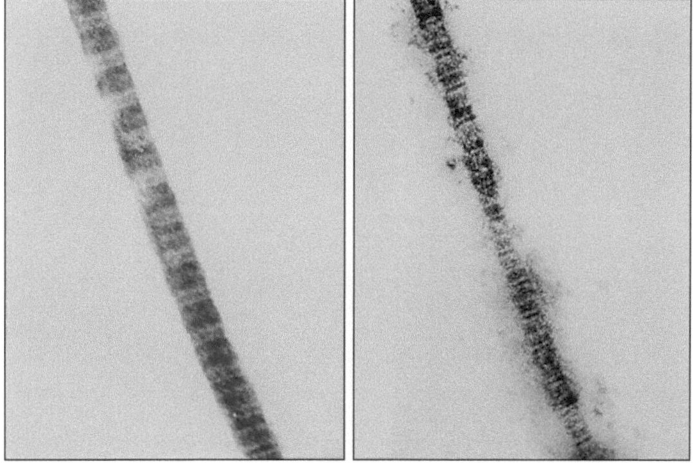

Fig. 25.3 Collagen before and after implantation. (Reproduced with the permission of Rolland Siegel)

The Zilla paper summary: "If bioprosthesis development continues as during the last four decades, the discrepancy between the world's needs and industry's commitment will continue to widen. Young patients need a truly long-lasting valve prosthesis rather than pursuit of continuous stepped-up sales efforts of non-innovative, long established products of yesterday."

In early June of 2008, Peter Zilla and David Williams (the editor of *Biomaterials*) paid Medical Carbon Research Institute a visit at its facilities on Cameron Road in Austin, Texas. Zilla had already implanted On-X valves, including at least one double, and expressed some concern with the implantability of double On-X valves. So we addressed his problem as described in the following chapter.

Reference

1. Zilla P, Brink J, Human P, Bezuidenhout D. Prosthetic heart valves: catering for the few. Biomaterials. 2008;29(4):385–406.

Chapter 26
On-X Valve Implantability

Zilla's primary interest, dissuaded by bioprostheses, was that he wanted to explore the implantation of On-X valves with a minimally invasive procedure. Using a catheter-delivered approach seemed impossible. More recently, an increasing number of surgeons are implanting On-X aortic valves using advanced minimally invasive procedures. Vinay Badhwar's EXPERT OPINION is the recent report on the first human experience, with an entirely robotic surgical aortic valve replacement (rAVR), using conventional prostheses. It was performed via a minimally invasive lateral thoracotomy working incision. Badhwar and his colleagues reviewed the technical aspects, early results, and potential for rAVR's future role in the management of aortic valve disease [1]. An image of the robotic implantation of a 19 mm On-X mechanical prosthesis is shown in Fig. 26.1. The paper has a video clip available online.

We responded to Zilla's expression of a problem in double valve replacement, saying the perceived problem had been addressed. A device to improve implantability was designed. A patent for the new introducer (Fig. 26.2) [2] was filed in 2006.

There had been several other ideas to improve implantability. Most were akin to the "shoehorn" concept. The concern with that approach was that the device could transmit substantial leveraged force that could fracture the valve. The inventive step in the new patent, setting it apart from a shoehorn concept, was attaching the device so it would be integral to the valve holder. The source of implant difficulties was coming from the flared inlet that needed to intrude inter-annularly into the aortic root.

While pursuing the implantation problem, I was pleasantly surprised by a long-distance dinnertime call from Sonja Van Riet, a heart surgeon working with Mervyn Williams after she had started performing On-X implantations. I was gratified to hear her tell me, "Dr. Bokros, thank you for the On-X valve!"

Surprised, I inquired why? She said, "It's so easy to implant!"

I had only heard about difficulties caused by the flared inlet, most originating from habits taught to cardiothoracic surgeons to "put in the largest valve you can."

J. Bokros, *Heart of Carbon*, https://doi.org/10.1007/978-3-031-17933-4_26

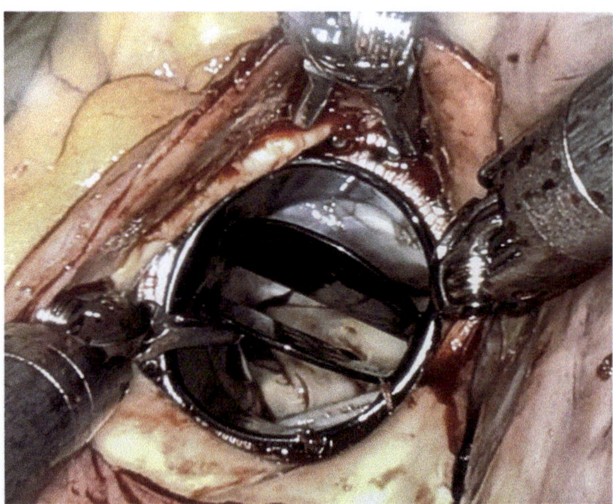

Fig. 26.1 A 19 mm On-X aortic valve after robotic implantation. (Reprinted from the *Journal of Thoracic and Cardiovascular Surgery*. Robotic aortic valve replacement, Badhwar V, et al., 2021;161(5) with permission from Elsevier)

Puzzled, I asked her to elaborate. She explained, "My mentor, Mervyn Williams, told me not to force a valve into place because traumatizing the tissue would exacerbate pannus overgrowth."

Looking at the size distribution of the valves implanted by Sonja, it was clear the distribution had shifted toward the small sizes, making implantation easier. Not to worry: even the 19 mm On-X valve provides mean gradients in the single digits, which is entirely normal.

What's in a Shape?

In December 2009, I finally saw a detailed proposal from Ruyra-Baliarda, MD, FETCS, which had been circulated within the company for nearly a year. His proposal was profoundly simple; his arguments were sound. It built upon the assertion made by Francis Fontan (Fig. 26.3) in 1994 that the sewing cuff should be shaped like the scalloped natural aortic valve. From our vantage point back in 1994, the scallops seemed too high. So we took the convenient path and made the sewing cuff circular, which I admit was a mistake.

Now however, a decade later and having gained an understanding of the importance of preserving the shape of the sinuses of Valsalva to maintain normal flow in the sinuses, the importance of providing a scalloped sewing cuff was back on the front burner.

The reaction of the VC-CEO to this shift was a blank stare. Even John Ely would not yield in his opposition to Ruyra's proposal. My persistence provoked a terse rejoinder from the VC-CEO: "We're too busy. Feel free to do it yourself."

United States Patent

Bokros et al.

Patent No.: **US 8,303,652 B2**

Date of Patent: **Nov. 6, 2012**

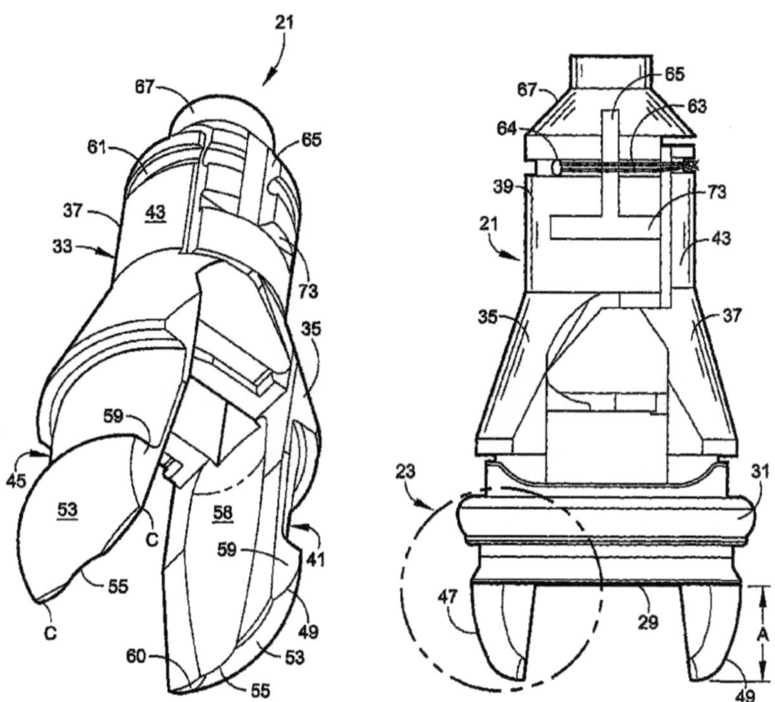

Fig. 26.2 US Patent No 8,303,652 B2, Heart Valve Inserter, issued November 6, 2012

So I did.

I called John Wright, an old friend with Genesee BioMedical in Denver, engaged in making annular rings and sewing cuffs. I explained to John what we needed.

The next week, I flew to Denver. When I arrived at Genesee BioMedical, a scalloped cuff was on the table. Not perfect—too tall. But soon we had a scalloped sewing cuff. The first prototype was a two-piece affair. Granted, two pieces were not acceptable for an implantable valve, but they were more than adequate for a rudimentary prototype to be tested in a wet lab for fit.

With our illustrious executives looking on, Regina Burnett, our VP of International Sales, whispered to me, "I'll bring Dr. Ruyra here to prove feasibility." *Great idea*, I thought. Regina had started with Carbon Implants as a secretary and, gifted with common sense, had advanced to VP of International Sales.

Regina arranged for Josep Gatell, the On-X distributor for Spain, and Ruyra to come to Austin for the occasion.

The wet lab served its purpose. The scallop of the sewing cuff matched the scallops of excised leaflet musculature. The reference points for orientation were the

Fig. 26.3 Francis Fontan discussing sewing cuff configuration. (Used with the permission of Jack Bokros)

nadirs, the three lowest points of the cusps, which had to be tied down first. With all the sutures in place, the valve could be lowered to the point where the individual nadirs coincided. It was a natural fit.

The prototype was a two-part assembly of a simple fabric covered orifice housing and a second scalloped portion that had to be stitched together. The scalloped portion was formed from a small-diameter Dacron tube, fashioned into a circle by suturing the two ends together. Then, the continuous tubular cuff was shaped using a scalloped fixture. Using steam heat, a perfect scalloped sewing cuff device was formed from a Dacron tube. The scalloped cuff had to be sutured to the orifice.

As noted above, this was suitable for an early prototype, but unacceptable for a clinical implant. The same sort of cuff would have to be made starting with a single tube of Teflon, with a diameter equal to the orifice of the sewing cuff.

Producing a scalloped orifice using a single tube of Teflon was easier said than done, so I presented the problem to Elva Correa, the lead in sewing cuff production.

Elva worked overtime and came in proudly the next morning to display a scalloped sewing cuff constructed using a Teflon tube. She smiled with a grin, as I had studied her approach just the day before and had concluded, "It won't work!"

Elva demonstrated how she sewed, stretched, and sewed and stretched, achieving a beautiful scalloped sewing cuff. What made it possible was the compliance of the Teflon fabric. Everything else fell into place as a consequence.

United States Patent
Ruyra-Baliarda et al.

Patent No.: **US 9,314,333 B2**

Date of Patent: **Apr. 19, 2016**

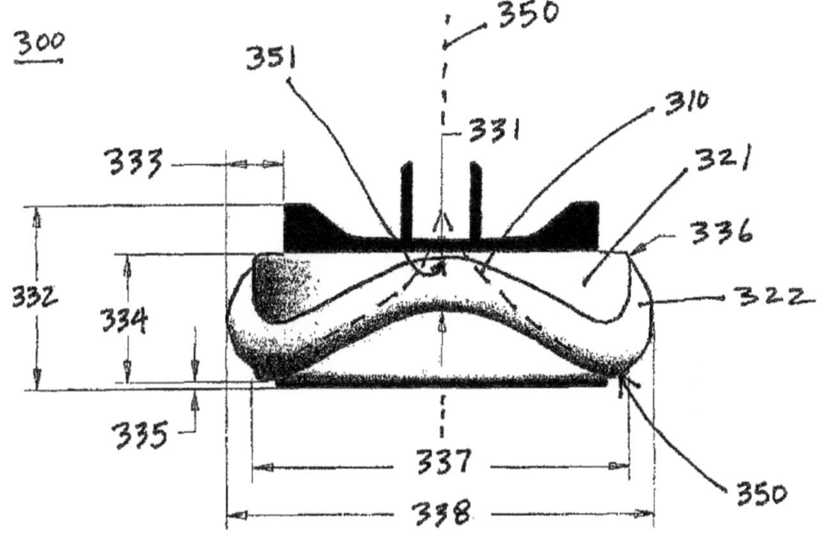

Fig. 26.4 US Patent 9,314,333 B2, Heart Valve Sewing Cuff, issued April 19, 2016. A fabric sewing cuff that consists of a cylindrical portion wrapped by a scalloped portion

In 2012, the sewing cuff was made available. When Fontan saw a photo, he responded with "Good idea."

He was too much of a gentleman to grumble "It's about time!"

In 2011, the concept was reduced to practice, with a patent filed on May 24, 2012. The patent was finally issued on April 19, 2016 (Fig. 26.4) [3].

References

1. Badhwar V, Wei LM, Cook CC, et al. Robotic aortic valve replacement. J Thorac Cardiovasc Surg. 2021;161(5):1753–9.
2. Bokros JC, Stupka JC, Waits CT. US Patent No 8,303,652 B2, Heart valve inserter, issued November 6, 2012.
3. Ruyra-Baliarda FJ, Gatell J, Ferre J, Southard FD, Bokros JC, Correa E, Poehlmann J. U.S. Patent No 9,314,333 B2, Heart Valve Sewing Cuff, issued April 19, 2016.

Chapter 27
Selling the On-X Valve

As the PROACT moved ahead, my time was devoted exclusively to promoting the On-X valve with the United States distributors and, following Regina Burnett's lead, worldwide distribution as well.

Regina served as our VP of International Sales. She and I set up places to tour the world, promoting the On-X valve as the only valve replacement that preserves normal flow.

Regina told me that our first jaunt would be to Syria because Syria had a very active heart valve replacement program. I was prepared and jumped on it. At the time, I was providing reprints of important publications to every person attending my presentations. Many were not fluent in English. I prepared to leave in early March and spend a couple of weeks there.

I arrived in Damascus on March 11, 2010. Regina's distributor was eminently professional. His staff members were also knowledgeable professionals, one a second-generation engineer.

The day after arrival, we motored to Aleppo for a trial presentation: a 40-min slide show. There were a couple dozen attendees: surgeons, cardiologists, and a number of listeners from the general medical field. I was impressed by their attentiveness and insightful questions. Everything felt quite positive.

After several local stops in Aleppo, we motored back to Damascus. I could not help but notice a lot of activity. I was told, "We're preparing for war." I did not follow up on that point, but when we stopped for lunch, the restaurant facility—which had the capacity for large gatherings—was nearly empty.

We sat down in a beautiful restaurant to an interesting Syrian lunch. After our meal, out came the exotic bubble apparatus where you "breath in, hold it, and then slowly exhale." When I tried it, I was assured that no drugs were involved. A small side chamber of the bubble apparatus was designed for generating smoke—smoke from anything, including drugs.

Back in Damascus, I was put up at a large luxury hotel. It reminded me of similar accommodations in Hong Kong. Eighty persons received invitations. The occasion

© The Author(s), under exclusive license to Springer Nature Switzerland AG 2023
J. Bokros, *Heart of Carbon*, https://doi.org/10.1007/978-3-031-17933-4_27

was widely promoted. The room for the presentation was large. One side was theater-like and the other side set up as a formal dining area. I was duly impressed.

On the day of the meeting, I was met at the entrance by women in formal dress. I could have been in Hong Kong or even Shanghai, but there were no Chinese present: only Syrians in Western attire ushering and greeting attendees.

The theater-like side of the meeting room was configured like a studio. A professional video camera recorded everything. I was informed that about half of the RSVPs were received. Each person received a folder with the important papers I would reference in my presentation.

The presentation went smoothly. The discussion was lively and everything was recorded for future use.

I was informed by our distributor that the persons invited who did not attend would be invited to a later presentation to view the recorded video, to be followed by a nice dinner.

I returned with two large empty suitcases, fully gratified with my Syrian experience but vowing to provide attendees with flash drives in the future in lieu of hauling bulky reprints across the globe.

Subsequent feedback from Regina revealed that the Syrians were implanting a lot of On-X valves.

While this sales effort was progressing, unbeknownst to me, the CV-CEO asked Siegel to execute a new consulting agreement. Later in 2010, Siegel called to tell me he was terminating the new agreement as he was not being utilized.

So, Siegel was lost. Four decades of marketing and sales of heart valve replacement experience not utilized—disappointing! But I took a deep breath and resumed my work.

As summer was fading in 2010, Regina had my time fully occupied. I wasn't dismayed by having so full a plate. When the rationale is clear and simple, "Natural by Design" presentations are a joy to deliver. As long as I was given free rein, I was happy, although I still missed Siegel.

My itinerary over the next 5 years is telescoped below, referencing only the geographic regions we covered rather than individual venues.

2010
October 23 to 31	Beijing, China
October 31 to November 6	Seoul, Korea
November 6 to 12	Shanghai, China
November 25 to 28	Helsinki, Finland
November 29 to December 4	Stockholm, Sweden; Oslo, Norway
December 5 to 8	Copenhagen, Denmark

2011
June 17 to 28	Hanoi and Saigon, Viet Nam
June 29 to July 4	Tokyo and Hiroshima, Japan
August 15 to 24	Tokyo, Japan
September 21 to Oct 4	Lisbon, Portugal
October 4 to 12	Barcelona, Spain
November 26 to December 3	Guadalajara and Monterrey, Mexico

2012
  March 22 to April 7   Europe
  August 23 to September 10   Tokyo, Japan,
                                               Hong Kong, China,
                                               Kuala Lumpur, Indonesia
  September 18 to 30   Europe: Budapest, Bulgaria, Kiev,
                                               Riga
  October 21 to 31   Paris, France, and Barcelona, Spain

2013
  January 12 to 26   Ireland, Amsterdam, London, Paris
  Winter storm in Europe meant I slept on the floor of a hotel in London as flights to Paris
  were canceled. I took the "Chunnel" train from London to Paris to meet with our French
  distributor. Then I returned to Texas
  February 8 to 23   Saudi Arabia
  March 17 to 25   Seoul, Korea
  April 27 to May 18   Australia and New Zealand
  August 26 to September 3   Seoul, Korea
  October 3 to 8   Vienna, Austria
  November 29 to December 6   Rome, Italy
  —rerouted return to Austin due to ice storm

2014
  March 31 to April 8   Istanbul, Turkey
  April 9 to 10   Glasgow, Scotland
  October 10 to 16   Milan, Italy

Late in 2012, 6 years from the inception of PROACT, valve sales were growing slowly, by no means reflecting the fact that we had achieved our primary design objective from 1994, viz., "Natural by Design."

I repeatedly told On-XLTI leadership, i.e., the venture capital leadership through their VC-CEO, that the market plan must first detail the design—the long (natural) valve length with a flared inlet and an actuated pivot that allows the leaflets to open fully and close reliably. Next, pose and then answer the question, "What is natural flow?" It's a natural spiral streamlined flow, initiated in the left ventricle, enabled by the valve design to progress without disruption through the valve into the coronaries.

Normality, and the fact that On-X carbon itself does not elicit thrombotic complications, makes the move toward anticoagulation reduction reasonable.

Their response was confidence that a first-of-its-kind IDE for low-dose anticoagulation had been approved and was sufficient to convince a skeptical market. I didn't agree because I understood that surgeons had been very skeptical of changing the valve they have been implanting to a new valve.

My daughter Kathy was employed at Medical Carbon Research Institute as a graphic artist. She had a degree in biochemistry but wasn't happy with laboratory work. She loved working as an accomplished graphic artist.

Kathy constructed a set of 28 slides demonstrating the natural design of the On-X valve. The series was my first big introduction to the world of design, the domain responsible for visually shaping my presentations.

Encountering Unexpected Resistance

None of the On-XLTI executives would use the animated 28-slide presentation. Instead, they used the single graphic enumerating the seven bullet point features of the On-X valve, trusting that quick-and-dirty was good enough.

I knew they were wrong: those bullet points were confusing and inadequate as a marketing tool. To the executives, however, the 28-slide presentation (which was actually informative) was simply too complicated to bother with.

The new presentation was a rapid-fire series of images that appeared movie-like, depicting the animated morphing of the St. Jude pivot to become the On-X pivot, allowing the leaflet to open completely. Its actuation assured reliable closure from the full open position (Chap. 20). Presented properly, it was a joy to watch. If practiced, it took 2–3 min to present, providing a marvelously clear visualization of valve dynamic fluid flow.

The animation of the opening and closing of the leaflets was worth a thousand words.

Frustrated by the studied avoidance of the slideshow by the executives, I wrote a script to use with the slides and distributed both the slides and script to the On-X executives and every single distributor. A memo was promptly issued, informing me that prior to *any* dissemination, all proposed marketing materials "must be approved by the VC-CEO."

Then I received a call from the VC-CEO informing me that Kathy had resigned.

"We had a disagreement," he claimed.

I knew she did not tolerate nor appreciate the impoverished marketing pieces being done by others and resented having her work being routinely gutted to "improve" it.

The VC-CEO entered my office, closed the door, sat down, and told me that the Board had told him that I was his most valuable asset and to inform me that I could continue as long as I liked. At 81 years of age, I was surprised and pleased with this news. My motivation was to promote the best of all possible valve replacements and to course-correct the inexplicably timid On-X marketing plan. I was admittedly critical of On-XLTI's marketing efforts and was hell-bent to cure the underlying problem. I had no plans to leave Medical Carbon Research Institute until that job was done.

By the end of 2012, I was becoming increasingly outspoken, perhaps even annoyingly so, in my attempts to influence our marketing strategy. I never stopped trying to persuade the marketing arm to openly leverage our valve's superior design and promote the normal flow it produces.

There was no traction to be gained by sticking with the anticlimactic claim that we had "the only valve to ever receive an IDE for a low intensity anticoagulation." That marketing angle amounted to a big nothing-burger compared to the much more compelling, industry-shaking story that kept being derailed, apparently never to see the light of day.

We simply weren't putting our best foot forward. And to hamstring our marketing effort in that way made no sense to me whatsoever.

In December 2012, I sold my pecan orchard and was looking for a building site for a new home. As chance would have it, my daughter who had resigned moved to a remote subdivision in Texas. I had visited that area in 1980 when CarboMedics, Inc. first moved to Austin. It was remote. The freeway had not yet been developed. But in 2012, the area was accessible, and I bought a home-site next to hers.

After our house was built, Kathy's interest in getting the On-X valve story out was rekindled, and she plunged in with both feet, helping me develop the required graphic materials.

I told her about my continuing frustration with the corporate promotion of the On-X valve. When I inquired of the VC-CEO why the PROACT "seemed to be stalled," the response was, in my opinion, more cryptic than I expected (i.e., some political pushback mixed with laments that "it's really tough").

Chapter 28
The Leonardo da Vinci Video

I told Kathy that an upcoming European Association of Cardio-Thoracic Surgeons meeting was to be held in October, 2014, in Italy. Further, the marketing manager advised me that there was a da Vinci meeting room near the venue, ideal for a presentation with a Leonardo da Vinci theme.

It was a perfect venue for a presentation focused on the On-X design. "Unique among all others, it preserves normal flow into the sinuses." The da Vinci theme would be the draw.

The marketing manager, with my support, convinced the VC-CEO to reserve the da Vinci room. He agreed that I prepare a presentation to be introduced by pointing out da Vinci's pioneering contributions to understanding blood flow through the aortic valve. We'd make a short introductory video to preface my verbal presentation.

The VC-CEO thought that the average surgeon's attention span was only 3 min which I challenged, asserting that if you could not keep a surgeon's attention for 3 min, you probably shouldn't be a salesman.

He pointed out another restriction: there was only $6000 in the budget for such a video. I responded that I'd personally contribute $3000. He agreed to this arrangement. When we started laboriously producing the video, Kathy worked in her home studio to generate the requisite graphics.

At start-up planning for the video, I informed the VC-CEO that Kathy was doing my graphics. He stood up dramatically and bluntly informed me "I can't work with her."

I responded, "She's not working for you. She's volunteered to do it—and is quite happy to do it. She won't let me pay her."

With a scowl, he sat down and directed that it would be the marketing manager's project, which was logical because the marketing manager had already made a number of videos and had already established an important vendor relationship of which I was unaware. The VC-CEO warned the marketing manager that I had to be happy with the final result. Moving forward, Kathy and I developed a script, along with Kathy's illustrations, for the marketing manager to edit.

J. Bokros, *Heart of Carbon*, https://doi.org/10.1007/978-3-031-17933-4_28

I had already found two published volumes, one an Artabras book [1] and the other, *Leonardo da Vinci on the Human Body* by Charles O'Malley and J.B. de C.M. Saunders [2]. These volumes provided all the background material I would need.

I realized early on that assimilating and understanding da Vinci would not be a quick study. After reading for several weeks, I realized I needed to shift gears and keep the focus on the anatomy of the heart.

The itch to discover da Vinci can be satisfied by reading the Artabras book [1]. It's derived from the 1938 da Vinci Exposition held in Milan. Flip over to *Leonardo's Optics* on page 405. If you start reading the first two pages of this section, chances are you won't close this volume for quite a while.

Late in life, circa 1510, Leonardo and his entourage were invited by Pope Leo X to the Belvedere, located at the highest point on the Vatican property. It was a "workshop" where famous people from the world could gather to pursue their own interests for extended lengths of time.

Around 1513, da Vinci turned his attention to anatomy, dissecting the heart of an ox. The ox was chosen because the Pope had forbidden human dissections at the time.

Employing his extraordinary dissection skills and remarkable visual perception, da Vinci sketched the geometric arrangement and described with incredible precision how blood flowed from the heart's aortic root into the sinuses of Valsalva (Fig. 28.1a–c). Leonardo's words were the first ever to describe normal blood flow through a heart. He surmised that vortical flow assisted the closure of the aortic leaflets.

It took about 500 years thereafter for Francisco (Paco) Torrent-Guasp, a Spanish scientist, to identify the HVMB. Then, 50 years later, Buckberg and his associates elucidated the sequential contraction of the segments of the HVMB to generate a helical ejection from the heart through the aortic valve [3] (Chap. 24).

In 1994 (Chap. 20), Medical Carbon Research Institute designed a prosthesis that preserved natural flow emanating from the left ventricle, the first valve ever to

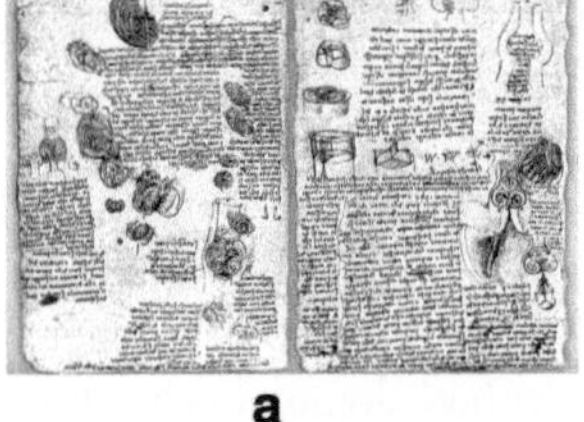

a   **b**   **c**

Fig. 28.1 (**a**) Leonardo da Vinci's actual journal pages showing his precise description in Italian of the blood flow through the aortic valve into the heart. It was unique to Leonardo da Vinci to write a mirror image of Italian. (**b**) An English translation of Leonardo's mirror-image Italian. (**c**) A paraphrase of the English translation of Leonardo's mirror-image Italian. (Reproduced with the permission of Kathleen Selbrede)

preserve normal flow without turbulence as described by da Vinci. That was the primary goal.

Kathy and I developed the introductory 12-min video, starting with da Vinci's dissection of the heart of an ox. After the video, I would go directly into another 20–30 min introducing the design, function, and flow of the On-X valve, leaving about 15–20 min at the close for questions.

The video was in draft form for several months before the meeting in Milan. The VC-CEO had all the executives of On-XLTI and a number of others view and approve it. When I got a copy, my original draft and some of the video graphics had been altered, and several unacceptable graphics had been added.

The original draft had included all the video graphics numbered and filed. I would not approve the marketing manager's version and documented the reasons why. There were words, me to the VC-CEO. Finally, the VC-CEO said we could proceed to develop the video on our own.

The marketing manager, however, refused to name the video vendor. "Get your own vendor." If that was to be, I told the VC-CEO, it wouldn't make sense because the marketing manager's vendor was familiar with On-X and had many of the graphic assets already on file. A new vendor would be more expensive and getting them up to speed "would not make business sense."

The VC-CEO advised that we were free to go forward on our own with the marketing manager's vendor, whom he identified.

It turned out to be a very fruitful collaboration. Kathy would bring her computer and work directly with the vendor. They both spoke the same "language," the jargon of those skilled in computerized graphics. The process was a breakthrough in efficiency and cost, as Kathy was volunteering her time to the video project. She, too, wanted to see the On-X valve given a fair shake in the marketplace.

Marketing Philosophies in Conflict

Late in October, I received word from the marketing manager that the VC-CEO had cancelled the da Vinci room reservation. No reason was provided. The 12-min video was to be shown on a single 20 × 30 inch video monitor set up with four pairs of headphones at the back of the sales booth.

No wonder sales were lagging! As if to prove it possible to make things even less favorable for screening the video, marketing then situated the video monitor on the side and rear of the On-X booth, further undermining its potential impact. The "not invented here" specter loomed large in my mind.

The next important event was the American Association of Thoracic Surgeons 95th Annual Meeting to be held in April 2015 at the Washington State Convention Center in Seattle, WA. I attended the distributor meeting that preceded the opening of the annual meeting. The VC-CEO customarily used the first hour to do cheerleading. He seemed worried. I customarily did not get to see our sales reports, but it was soon clear that sales were lagging.

His opening slide showed a silver bullet, centered on the screen, about 4 feet long. "There's no silver bullet," the VC-CEO authoritatively proclaimed.

Struggling, he pivoted to a stalwart call for stiff upper lips: "We have to work hard and be patient. Sales will rise," he promised.

Then things got worse.

One of the large US distributors told me at the break that he needed me to come and present to one of the surgical groups. But he said, "You are persona non grata— my request for your visit was denied." My feeling about that was not good, but this news substantiated my growing suspicions. Had I been too candid in my criticism of the VC-CEO's marketing skills?

The next morning, I entered the American Association of Thoracic Surgeons Exhibit Area early. So did the VC-CEO. I made an oblique comment about the new booth. I said it was very pretty, but it made no mention of On-X's unique natural spiraling flow.

Several unflattering words later, he told me, "If you don't like it, take it to the board."

"It's too late for that," I pointed out.

On a number of occasions, the VC-CEO had, for his own edification, described On-XLTI as a turnaround situation. This characterization irritated me because by 2002 the FDA had already approved the On-X valve. On-XLTI was neither a turn-around nor a distressed asset. It was a true start-up.

Not mentioning my American Association of Thoracic Surgeons experience in Seattle, I shared the news with Kathy and mentioned that we should start by retaining some of the graphics from the Milan video. We would use only a few words to cover the da Vinci theme and then emphasize the valve design and the normal flow it produced.

I told her we needed to make a video that visually delineates normal vortical and helical flows and then embed visuals of actual natural flow using the works of prominent scientists involved in characterizing blood flow in humans. After that, we should demonstrate how the organized spiraling ejection proceeds from the left ventricle through the aortic valve into the sinuses without disruption.

In short, preserving natural flow was the key. Reducing anticoagulation intensity and attaining normal hemodynamics with minimal blood damage are the critical consequences of preserving natural flow.

I told the VC-CEO that we were changing the video to emphasize the design, demonstrating the elimination of turbulence and preservation of natural flow: "Natural by Design."

"I'm paying for it," I alerted him.

"Okay," he responded.

The video we were making was to decisively reveal the existence of the silver bullet that the VC-CEO did not think existed.

All the while, Kathy and her husband were copyrighting their work: Kathy on her graphics and Martin on his background music sound track.

References

1. Leonardo da Vinci, An Artabras Book. Reynal and Company in association with William Morrow and Company, New York. All rights reserved under International and Pan-American Copyright Conventions and copyright by Instituto Geografico De Agostini, Novara, Italy
2. Leonardo da Vinci on the Human Body by Charles D O'Malley and JB de CM Saunders. Wings Books, New York. Avenel Random House ISBN 0-517-38105-2
3. Coghlan C, Hoffman J. Eur J Cardiothoracic Surg Suppl. 2006;295:S4–S17.

Chapter 29
Gerald Buckberg Joins the Video Project

In January 2015, while attending the Society of Thoracic Surgeons meeting in San Diego, I spotted Professor Gerry Buckberg—a distinguished scientist and prominent contributor to our understanding of the form and function of the heart—roaming the exhibition area. We had met some years earlier at a lunch. We talked then about the On-X valve but mostly about Buckberg's early work.

I was amazed at Gerald's memory and was very pleased that he recalled our meeting. He immediately posed a question, all but treating me as a student.

"At the end of systole, when does the mitral valve open?" he asked.

I thought it was obvious that it opened after the aortic valve closed. I sensed that it was a trick question and replied, "It opens before the aortic valve closes."

His eyes opened wide. "You're right," he remarked.

If he'd asked me why I answered as I did, however, I would have drawn a blank.

Instead of asking me why, he launched into the full explanation.

"At the very end of systole, after the descending segment has stopped contracting, the ascending segment continues contracting creating a hiatus as the heart muscle rotates, pulling the mitral cords, which opens the valve early."

I don't remember my precise response, but I did pivot the conversation to the video Kathy and I were making. I related that she was doing the graphic arts and Martin, her husband, was providing the symphonic background music. Once Gerald grasped that this was "a family project," whenever I asked for his help or counsel about the video thereafter, he graciously made time for me.

By May, 2015, with the help of Buckberg, we'd added two more important researchers prominent in the world of four-dimensional magnetic resonance imagery. One was Philip Kilner, working with Magdi Yacoub's group at the Royal Brompton Hospital in London. The other was Michael Markl, working at Northwestern University. As we were developing the video, Buckberg, Kilner, Barker, and Markl were linked into our ongoing email dialogue.

In our first email dialogue concerning the On-X valve, Kilner provided two references pointing to the early work by Bellhouse [1, 2]. In his letter on the function of

J. Bokros, *Heart of Carbon*, https://doi.org/10.1007/978-3-031-17933-4_29

the mitral valve to *Nature*, Bellhouse had used his experimental wizardry to demonstrate that the filling of the ventricle must be rapid, not slow.

This finding would explain why the mitral valve opens prior to aortic valve closure, as Gerry Buckberg had shown to be true. Nature indeed provides the mechanism that Buckberg has described. The mitral valve is not opened due to pressure differential, but rather as a result of the concomitant rotational motion of the heart with its papillary muscles, caused by the final shortening of the ascending sequence of the HVMB. The heart opens the mitral valve early in the last phase of systole. This I find absolutely fascinating.

Buckberg's input prompted our modification of the Wigger diagram used in the video. His further insistence that "lines in nature are not straight" led us to correct what we had previously depicted in our explanatory animations.

We first tried explaining to Buckberg that the flow through the aortic valve, though appearing straight, had curvilinearity with a radius of curvature approaching infinity. His emailed response amounted to an annoyed "Get real!" He was not satisfied by that rationalization. Once we identified our lines as being merely schematic in nature though, we were able to get past that sticking point with our vigilant counselor.

On another subject, Buckberg emailed us to say that engineers who don't follow nature's lead are destined to fail. The focus of this maxim was directed to our use of a bileaflet aortic concept (a variant not found in nature) rather than a trileaflet design.

I explained that if an imaginary elf stood at the apex of the heart looking up, he would see a smooth funnel shape leading up to the aortic valve. When the natural leaflets opened, the funnel shape continued into the sinuses without disruption. If the valve were trileaflet in design, the elf would see the leaflets pivot open, leaving three gaps to distribute the flow, transforming one smooth flow into three peripheral channels and a fourth central jet.

While the thin flat leaflets of a bileaflet valve opened unnaturally, in mid-stream, they'd blend in with the streamlined flow to preserve natural, non-turbulent flow.

Buckberg responded, "Okay."

The trileaflet concept departs from nature but the bileaflet preserves normality.

Buckberg quipped, "I think Mother Nature would agree."

Misconceptions That Could Have Been Prevented

In early 2015, the VC-CEO inquired how the new video was coming along. I told him the first 6 minutes were complete. The next day, I played it for him all the way through the soft landing. He was impressed about the soft-landing depiction and proposed that the soft-landing part should be published.

I was surprised and immediately reminded the VC-CEO that in 2011 Larry Scotten and Rollie Siegel had published just such a paper comparing transient closing impacts for mechanical and bioprostheses [4]. In the same year, they presented their data at John Ely's Investigator meeting.

He then abruptly changed the subject, informing me that the venture capitalists were getting impatient and were looking for a buyer for On-XLTI. I advised Kathy we needed to move fast, as the silver bullet needed to be available for rollout very, very soon.

In 2011, I had taken notice of the relevance of Scotten's publication and asked Ely to send Scotten some On-X valves to test. Test he did, and the results were mailed to On-XLTI in December of 2011. The results showed the On-X valve compared to the St. Jude Regent valve, St. Jude's best challenger to the On-X hemodynamics, measured as reverse backflow velocity in meters/second, was superior. For aortic valves, the values are about minus 80 m/s compared to about minus 212 m/s for the Regent. Smaller values are better. In the mitral position, the values are about minus 60 compared to minus 140 m/s. Again, the On-X valve's superiority to the St. Jude Regent is substantial. In fact, in this regard, On-X valves are equal to tissue valves for both the aortic and mitral positions. I incorporated those results within the video in 2015.

Big News from the FDA

Finally, after 10 years, word arrived that on April 1, 2015, the FDA had granted approval for On-X aortic valve high-risk patients to receive a reduced level of anti-coagulation. The INR limits were reduced to 1.5–2.0. With a sigh of relief, everyone involved was elated. Those remarkable results were published [3].

Later in May 2015, Philip Kilner expressed to Gerry Buckberg and me that he liked the On-X valve. But he still had a serious reservation: "All mechanical valves *slam* shut," he affirmed.

I responded by sending Kilner the 2011 Scotten data that showed that the closing geometry we had designed significantly reduced the mechanical impact of the On-X valve. All this important old news had gone undisclosed by On-XLTI, so I included it in the video to bring it current. This had been a marketing oversight (Reference Chap. 20 for a detailed explanation of soft landing) which allowed a serious misconception about the On-X valve's performance to persist unchallenged.

References

1. Bellhouse BJ, Bellhouse FH. Mechanism of closure of the aortic valve. Nature. 1968;217:86–7.
2. Bellhouse BJ, Bellhouse FH. Fluid mechanics of the mitral valve. Nature. 1969;224:615–6.
3. Puskas J, Gerdisch M, Nichols D. Reduced anticoagulation after mechanical aortic valve replacement: interim results from the prospective randomized on-X valve anticoagulation clinical trial randomized Food and Drug Administration investigational device exemption trial. J Thorac Cardiovasc Surg. 2014;147:1202–10.
4. Scotten LN, Siegel R. Importance of shear in prosthetic valve closure dynamics. J Heart Valve Dis. 2011;20:664–72.

Chapter 30
A Surprise from Northwestern University

A poster (Fig. 30.1) from the Northwestern group [1] was valuable not only for the data that it contained (defining the On-X valve's preservation of nature's normal flow) but because it referenced a recent publication from a pilot investigation from a group in Berlin (*International Journal of Cardiology* 2014) [2]. I was excited by the report, which struck me as possessing landmark status. The poster shows that blood flow through On-X aortic valves preserves natural flow. No other valve, mechanical or biological, preserves natural flow.

Just as I stated in Chap. 4 under Fig. 4.5 to eliminate the reader's confusion upfront regarding the bonding of heparin to carbon to be the source of the thromboresistance, in a similar way, for the Berlin groups' investigation, the word mechanical prosthetic valve doesn't characterize a valve's thromboresistance. Take notice upfront that the Berlin groups' investigation did not include On-X prosthetic valves, especially with respect to thromboresistance.

Focusing on remodeling of the ascending aorta after aortic valve replacement, the Berlin group applied advanced 4D-flow MRI technology to resolve the complex phenomenon of turbulence into measurable helical and vortical components. These metrics allowed one to calculate peak wall shear stresses during systole.

The Berlin team analyzed blood flow characteristics in the ascending aorta after aortic valve replacement. They had five groups of patients ($n = 47$).

For each group measured, the researchers mapped wall shear stresses over 12 segments along the aortic circumference, using 3 analytic planes oriented perpendicular to the aortic wall and positioned downstream of the valve in the ascending aorta (Fig. 30.2). Their goal was to evaluate and compare the eccentricity of the flow for a variety of aortic valve replacements (not including the On-X valve).

Eccentricity of the flow causes wall shear stresses that, in turn, can damage the aortic wall and induce a variety of pathologies. The images in Fig. 30.3 are composites of all of the valves in each group, so some image quality is lost as a result of the conflation process. The individual images, however, exhibit significantly higher resolution.

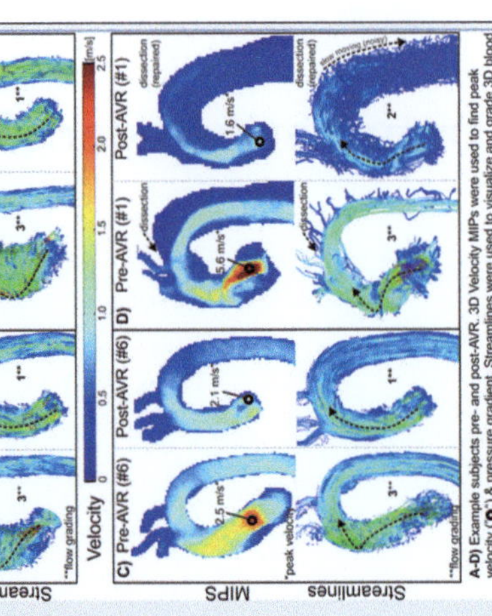

Fig. 30.1 Northwestern University Feinberg School of Medicine poster Proceedings of the Heart Valve Society Monaco 2015. (Reproduced with the permission of S Chris Malaisrie MD)

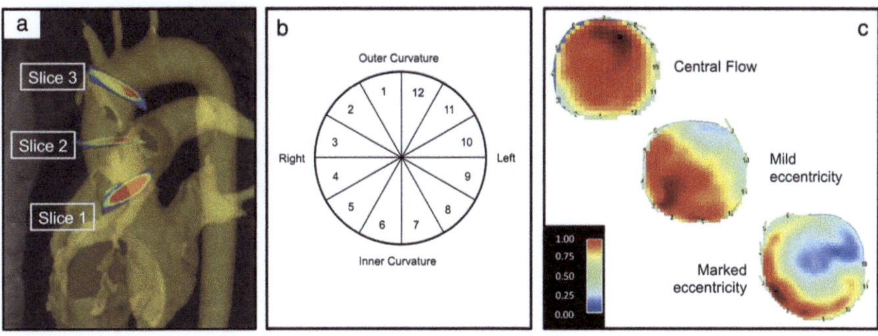

Fig. 30.2 (**a**) Position of the analysis planes in the ascending aorta: 1 the sinotubular junction, 2 mid-ascending, and 3 distal ascending aorta. (**b**) Distribution of analytical segments along the aortic wall circumference. (**c**) Peak systolic flow map at the level of the mid-ascending aorta demonstrating central systolic flow and mild and markedly eccentric systolic flow. (Reprinted from the *International Journal of Cardiology*. Blood flow characteristics in the ascending aorta after aortic valve replacement—a pilot study using 4D-flow MRI, Florian von Knobelsdorff-Brenkenhoff, et al., 2014;170(3) with permission from Elsevier)

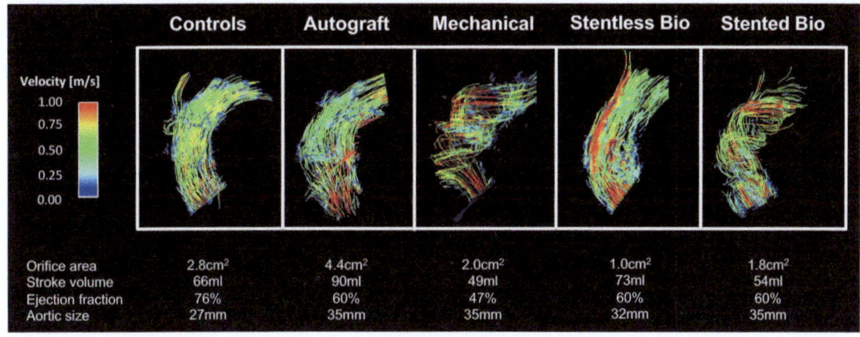

Fig. 30.3 Visualization of blood flow in the ascending aorta using particle traces during peak systole for each of the aortic valve replacements. (Reprinted from the *International Journal of Cardiology*. Blood flow characteristics in the ascending aorta after aortic valve replacement—a pilot study using 4D-flow MRI, Florian von Knobelsdorff-Brenkenhoff, et al., 2014;170(3) with permission from Elsevier)

The image on the left is representative of healthy control cases and represents the baseline for *normal*. The second image represents autografts which, as might be expected, are near *normal*. The third and fifth images are representative of *abnormal* flows that are departures from the baseline *normal* flow. Vortical, helical, and eccentric flows were defined and documented in order to compute an aggregate quantitative score for each group (Fig. 30.4).

The upper row shows the frequency of each score for the various AVR groups and controls. The lower row depicts the mean scoring results for each group.

The score for each valve group is shown. The overall grade of the group is the three values for vorticity, helicity, and eccentricity. For example, the respective

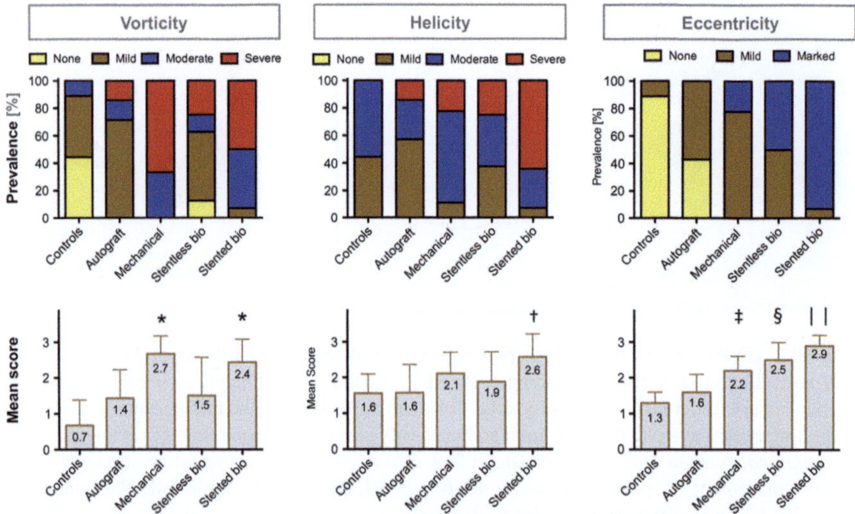

Fig. 30.4 Evaluation of vorticity, helicity, and eccentricity of blood flow. The upper row shows the frequency of each score group. The lower row depicts the mean ± SD scoring ($p* < 0.05$ vs stentless autografts and controls; $p + < 0.05$ vs stentless autografts; $p\ddagger < 0.05$ vs controls; $p\S < 0.05$ vs stented autografts controls; $p\parallel < 0.05$ vs stentless mechanical autografts and controls). (Reprinted from the *International Journal of Cardiology*. Blood flow characteristics in the ascending aorta after aortic valve replacement—a pilot study using 4D-flow MRI, Florian von Knobelsdorff-Brenkenhoff, et al., 2014;170(3) with permission from Elsevier)

scores for the mechanical group are 2.7, 2.1, and 2.2 for vorticity, helicity, and eccentricity. The numerical groups represent the grade and are labeled V H E.

The most *abnormal* vorticity, helicity, and eccentricity were for the mechanical valves and stented bioprostheses (Fig. 30.5). The corresponding values were 2.7, 2.1, and 2.2 and 2.4, 2.6, and 2.9, respectively.

The stentless bioprostheses, with scores of 1.5, 1.9, and 2.5 in Fig. 30.6, are an improvement over stented bioprostheses, but the elevated eccentricity is still indicative of *abnormal* flow.

For healthy controls, *normal* vorticity, helicity, and eccentricity were 0.7, 1.6, and 1.3. Corresponding scores for the autografts were 1.4, 1.6, and 1.6, respectively (Fig. 30.7).

On the left (Fig. 30.8a), baseline flow behavior (to use as a sharp contrast against the replacement valves) is established using a healthy volunteer control subject with mild helicity viewed parallel to the flow and marked by cohesive systolic streamlines and little vorticity viewed from the side.

On the right (Fig. 30.8b), two aortic valve replacements (one marked "Helix," a St. Jude Medical Regent mechanical prosthesis, and the other marked "Vortex," a Medtronic, Inc. Freestyle stentless bioprosthesis) are each graded as severe. The helical flow on the left is shown in a view parallel to flow through the sinotubular junction, whereas the vortical flow on the right is shown in a sagittal cut plane.

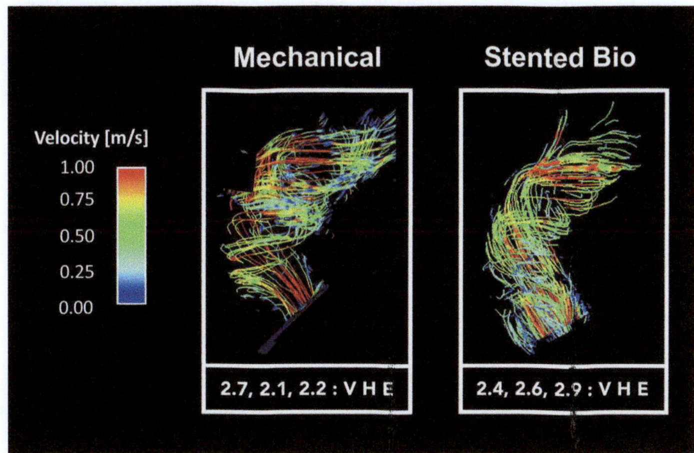

Fig. 30.5 Comparison of mechanical vs stented bioprostheses. (Reprinted from the *International Journal of Cardiology*. Blood flow characteristics in the ascending aorta after aortic valve replacement—a pilot study using 4D-flow MRI, Florian von Knobelsdorff-Brenkenhoff, et al., 2014;170(3) with permission from Elsevier)

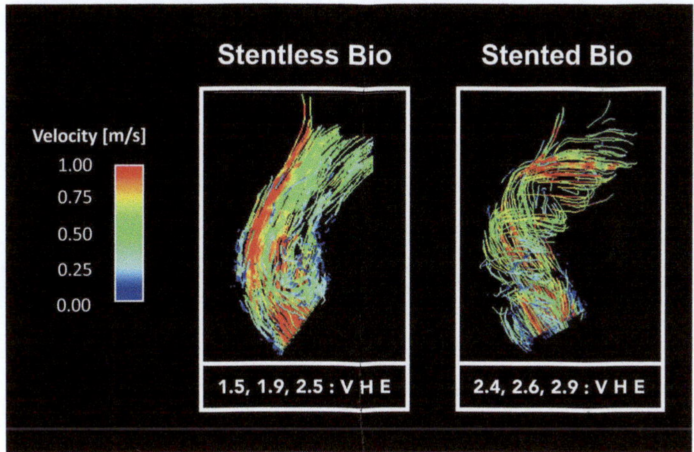

Fig. 30.6 Comparison of stentless vs stented bioprostheses. (Reprinted from the *International Journal of Cardiology*. Blood flow characteristics in the ascending aorta after aortic valve replacement—a pilot study using 4D-flow MRI, Florian von Knobelsdorff-Brenkenhoff, et al., 2014;170(3) with permission from Elsevier)

The Berlin study indeed represents a key contribution, perhaps a landmark, in the effort to understand *normal* versus *abnormal* flow. Although it was a pilot study, it points the way forward in respect to future investigation.

On the left in Fig. 30.9, the peak wall shear stresses for the three analysis planes in the ascending aorta are shown along the circumferences of the aortic wall for:

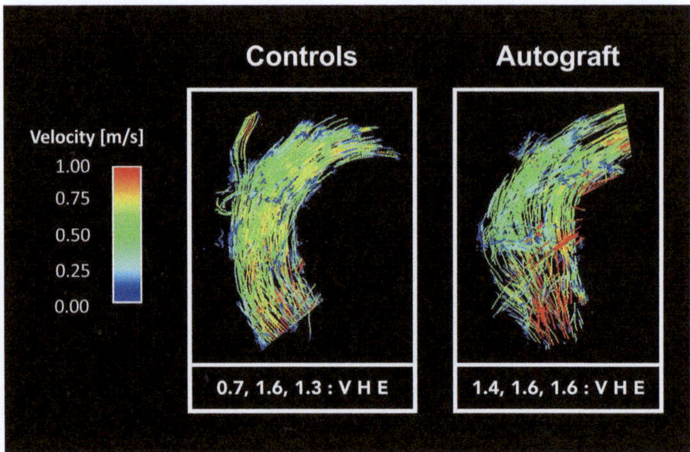

Fig. 30.7 Comparison of controls vs autografts. (Reprinted from the *International Journal of Cardiology*. Blood flow characteristics in the ascending aorta after aortic valve replacement—a pilot study using 4D-flow MRI, Florian von Knobelsdorff-Brenkenhoff, et al., 2014;170(3) with permission from Elsevier)

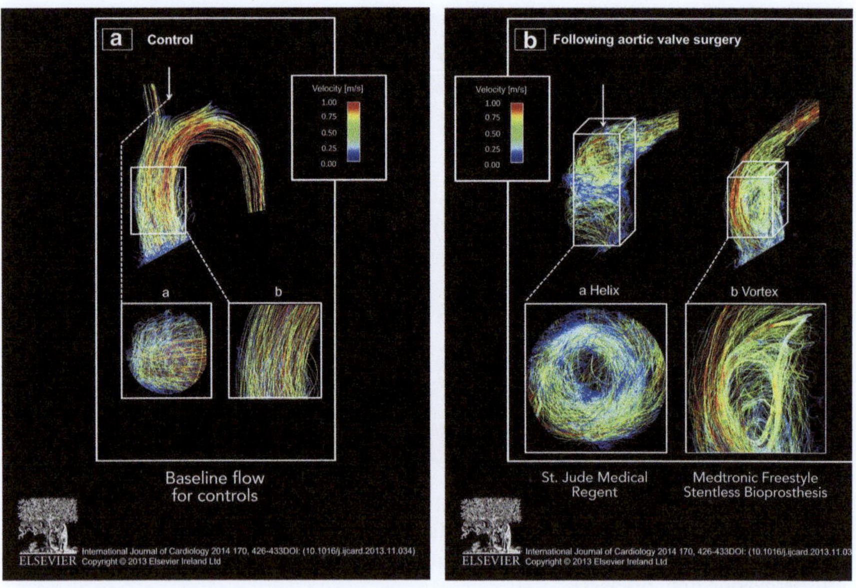

Fig. 30.8 (**a**) Control with mild helicity viewed parallel to flow and cohesive systolic streamlines and little vorticity viewed from the side. (**b**) Two aortic valve replacements one marked "Helix"(a St. Jude Medical Regent mechanical prosthesis) and the other marked "Vortex" (a Medtronic Freestyle stentless prosthesis) each graded severe. (Reprinted from the *International Journal of Cardiology*. Blood flow characteristics in the ascending aorta after aortic valve replacement—a pilot study using 4D-flow MRI, Florian von Knobelsdorff-Brenkenhoff, et al., 2014;170(3) with permission from Elsevier)

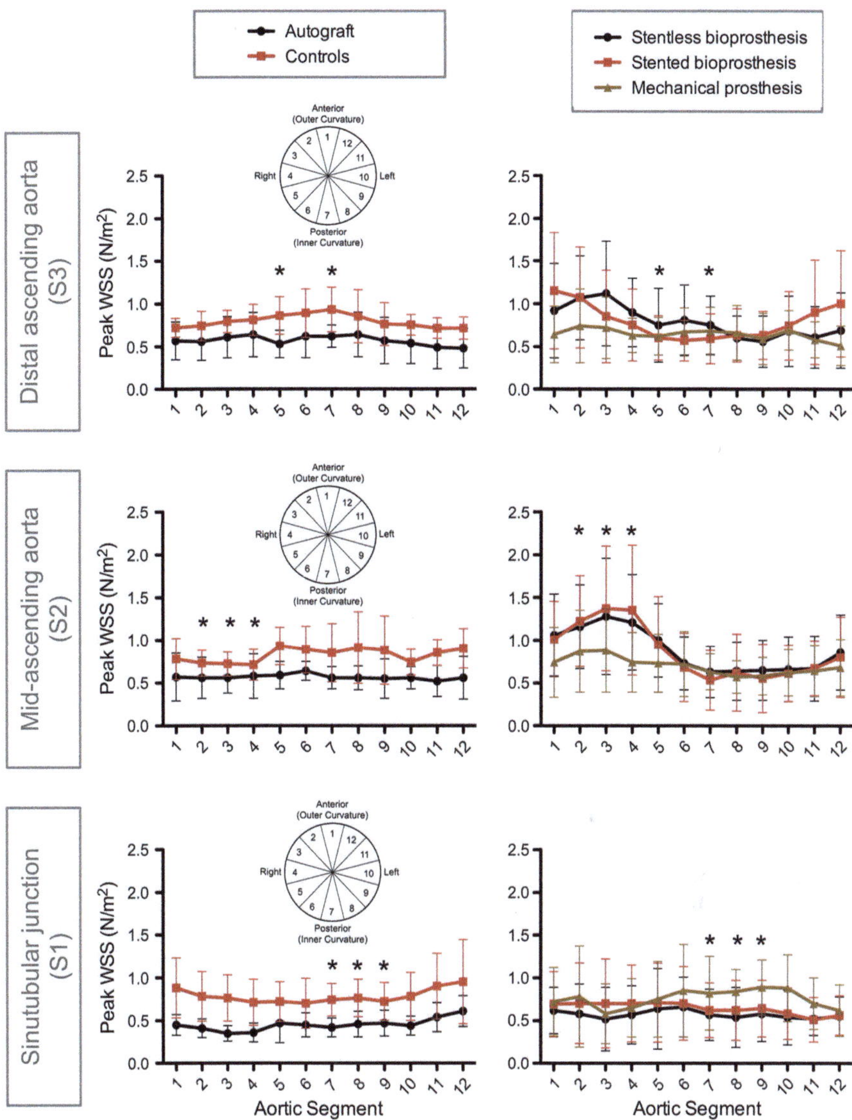

Fig. 30.9 Segmental distribution of peak wall shear stresses (WSS peak) along the circumference of the aortic wall for aortic levels S1, S2, and S3 for all groups of aortic valve replacement and controls. Asterisk indicates the presence of any significant inter-group difference at the given aortic location. (Reprinted from the *International Journal of Cardiology*. Blood flow characteristics in the ascending aorta after aortic valve replacement—a pilot study using 4D-flow MRI, Florian von Knobelsdorff-Brenkenhoff, et al., 2014;170(3) with permission from Elsevier)

1. The controls.
2. The autografts.

On the right, the peak wall shear stresses for the corresponding three analysis planes are shown for:

1. The stented bioprostheses.
2. The stentless bioprostheses.
3. The mechanical prostheses.

The data reveal that wall shear stresses for both the bioprostheses on the right in the mid-ascending aorta location are markedly *abnormal*. Excessive wall shear stresses are considered pathologic.

The Berlin study for the first time provides an objective basis for evaluating deviations from *normal* flow that impact heart function, not only in the ascending aorta but also in the sinuses. The study reveals that all the mechanical valves and bioprostheses measured were incapable of achieving *normal* flow. Again, On-X valves were not included in the Berlin study.

The Video Continues to Evolve

In May 2015, the VC-CEO asked for an update on the video. I assembled all of the material that characterizes *normal* flow, including Kilner's 1993 [3] study using MRI techniques showing that healthy human hearts eject a cohesive spiral flow through the aortic valve to form a vortical flow in the sinuses that assists the closure of the aortic valve. It also included Kilner's 2000 letter to *Nature* [4] that depicted the flow through the heart of a 34-year-old man, a streamlined cohesive flow through the entire cardiac cycle. Both of Kilner's papers were detailed in Chap. 22.

I also included the Berlin group's 4D-flow MRI data which characterized turbulence in terms of its basic vorticity, helicity, and eccentricity elements, allowing the calculation of peak wall shear stresses during systole. The study included mechanical and biological valves but did not include the On-X valve.

Finally, we prepared an updated chart, *Triumph Over Turbulence*, to include results from the Berlin study [1] as well as the poster (Fig. 30.1) from Malaisrie [5] displayed in 2015 at the Monaco meeting of the Heart Valve Society.

The next day, I laid out all of the material for the VC-CEO to study. He watched attentively for a few minutes, then pushed back, apparently overwhelmed, and said "I don't understand. It's too complicated, and I don't have time. I've already rejected the chart *Triumph Over Turbulence*," he said, even though he had not studied the updated version.

He then informed me about a possible sale of On-XLTI to CryoLife. It was to be in complete confidence. He became upbeat. The CryoLife attorneys needed to interview me about the Medtronic, Inc. restriction (listed in Chap. 18) that limited any such sale to companies that had not been in the heart valve replacement field.

I was favorably inclined to a possible sale to CryoLife and freedom from venture capitalist control. Up until now, I had felt like "Jack in a box." I was especially pleased and relieved that the technology I'd been pursuing since 1964 might have unique properties for use in additional medical devices.

The interviews with the attorneys from both sides were deposition-like related to litigation but weren't conducted under oath. All this happened as Kathy and I were working diligently on the "silver bullet" video.

Our initial objective for the video was to demonstrate to the venture capitalist (and its VC-CEO) that the On-X valve was the only valve among all the rest, mechanical or biological, that preserves coherent *normal* flow ejected from a healthy human heart without disruption. The On-X allows vortical flow formation in the sinuses to not only assist in valve closure but also produce *normal* flow into the coronaries. The consequences were reduction in blood damage, allowing for less intense anticoagulation but also providing *normality* of hemodynamics. The On-X valve is a paradigm shift in heart valve replacement.

This was the core of my message.

I was not sure whether either On-XLTI or CryoLife recognized the true value of the On-X valve. The title of the Malaisrie poster says it all: *Restoration of Physiologic Flow in the Ascending Aorta After Aortic Valve Replacement*. That, of course, was the paramount objective in the quest for the ideal heart valve replacement all along. In that light, I considered the valuation that I had heard to be far too low.

I had kept Gerry Buckberg apprised of our progress all through the development of the video, but was surprised and pleased to receive this email from him in July 2015:

> Dear Jack.
> > I am not an expert on fluid dynamics, but understand invention.
> > Congratulations, as you found its secret. Imagers watch, yet you created a mirror of nature. Nature displayed the fluid dynamic end point, and you taught us that this is possible by a structural mechanism.
> > You found this pathway mechanically, and may thus rid the patients of both the bleeding and thrombotic complications of other mechanical valves, as well as the need to replace a prosthesis at an elderly and more risky age.
> A job well done.
> Best, Gerry.

I, of course, inserted the email into the video.

Too Little Too Late?

On August 28, 2015, I requested a private meeting with one of the board members, the head of the second venture capitalist group to invest in On-XLTI. I felt the company was undervalued in the current negotiation and wanted to speak about that.

A meeting was arranged. The board member requested his attorney be present. The meeting was scheduled for Monday, August 31, at 9 a.m. in his office. I was puzzled for the need of a lawyer.

Word came back on August 30 that the meeting was postponed and needed to be rescheduled. So we got back to finishing the video.

The poster by Malaisrie et al. [1] in the *Proceedings of the Heart Valve Society*, held in Monaco in 2015 (Fig. 30.1), sought to apply the Berlin study's new metrics to valves omitted from that landmark work by assessing the hemodynamic outcomes of an On-X aortic valve replacement. It is important to note that two of this new poster's authors were also authors of the Berlin study, so the nomenclature and metrics remained consistent across both studies.

After reviewing the Berlin study results, and noting the relationship of the Berlin study and the poster by Malaisrie, I conjectured that the authors of both documents realized that the absence of the On-X valve from the Berlin study was a serious omission. While the Berlin group paper was in process, the preliminary findings on the On-X valve from the Northwestern group were emerging. The motivation for the prompt display of the poster in Monaco shortly after the conclusion of the Berlin study may have been the realization that the continuum from the Berlin study (where all heart valve replacements displayed *abnormal* flow) needed to be completed by including On-X valves.

Data being developed by the Northwestern group indicated that the On-X valve was an exception. That study was proving that the On-X aortic valve preserved *normality*! But *normality* remained undefined.

The poster's objective (Fig. 30.10) was to assess the hemodynamic outcome of aortic valve replacement using a mechanical prosthesis designed with an inlet geometry intended to reduce *abnormal* flow in the valve outflow region.

<u>Objective</u>: To assess hemodynamic outcome of aortic valve replacement (AVR) using a mechanical prosthesis designed with an inlet geometry intended to reduce abnormal flow in the valve outflow region.

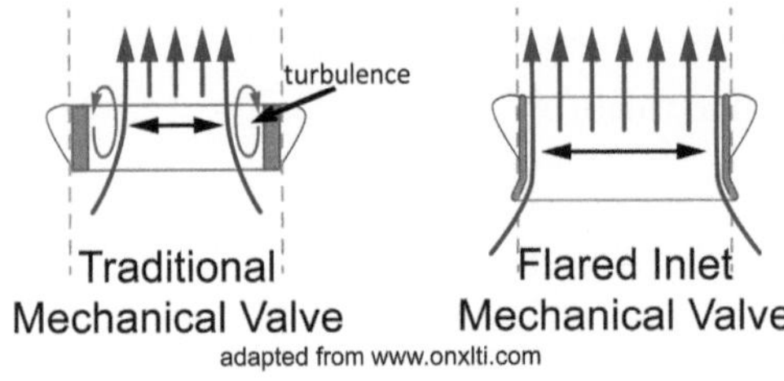

An abrupt change in the 'straight cylinder' inlet design (left) may promote turbulence and a decrease in the effective valve orifice area (EOA). A flared inlet may reduce turbulence & promote an optimized EOA.

<u>Thus, hemodynamic outcome was assessed by</u>:

1. Pre- and post AVR transvalvular pressure gradient
2. 3D blood flow pattern visualization using 4D flow MRI

Fig. 30.10 Objective. (Used with the permission of S Chris Malaisrie MD)

Results: Imaging, Demographics & Hemodynamics

				Pre AVR		Post AVR	
#	Age	Sex	Intervention	ΔP (mmHg)	Grading	ΔP (mmHg)	Grading
1	58	m	AVR+ARR	127	3	10	2
2	29	m	AVR+ARR	11	3	23	1
3	43	m	AVR+ARR	135	3	21	1
4	29	m	AVR+ARR	-	-	7	1
5	55	m	AVR+ARR	6	1	6	1
6	48	m	AVR+ARR	24	3	9	1
7	44	f	AVR	-	-	21	1
8	29	m	AVR+ARR	15	3	29	2
9	40	m	AVR+ARR	96	3	22	1
10	53	f	AVR	42	3	17	2
11	62	m	AVR+ARR	34	2	36	3
				54±48	**2.7±0.7**	**18±9**	**1.5±0.7**

Fig. 30.11 Patient demographics and hemodynamics. (Used with the permission of S Chris Malaisrie MD)

Hemodynamic outcome was assessed by pre- and post-AVR transvalvular pressure gradient and 3D blood flow pattern visualization using 4D-flow MRI.

Patient demographics and hemodynamics are provided in Fig. 30.11. All but 2 of the 11 patients were AVR plus aortic root replacement, labeled ARR. Two were isolated aortic valve replacements. The flow gradings prior to implantation were mostly 3. Post-implantation, they were mostly 1 and 2, averaging 1.5.

Evidence indicates that the flared inlet valve design has improved hemodynamic flows when compared to the previous report for the traditional valve designs used in the Berlin study, with mean gradings reported to be 2.7 versus 1.5 for patients receiving the On-X valve. Following valve replacement, flow was central, with minimal helicity or eccentricity, and the mean grading was 1.5, a value that falls within the range of 0.7–1.6 that represented *normality* according to the Berlin study.

The data included only On-X valve patients, whereas the Berlin study excluded On-X patients. A direct comparison with all others is yet to be performed.

Figure 30.12a–c is an example of clips from the 4D-flow MRI for an AVR patient from the poster before and after isolated implantation of an On-X valve. The first view shows the flow before the implantation, with a grading of 3. The second view depicts the flow after implantation. The flow is described as central, with minimal helicity, like the *normal* flow observed for the controls in the Berlin study, although *normality* is not yet precisely defined.

The third image (Fig. 30.12c) of the streamline ejection offers a glimpse of the retrograde flow back through the tiny gap at the tips of the two leaflets. The pivot purge that provides the hydraulic cushion is visible at the start of diastole.

The results of the Berlin study and the Malaisrie poster indicate that the On-X valve may have indeed achieved *normality*, meaning that the original hypothesis set

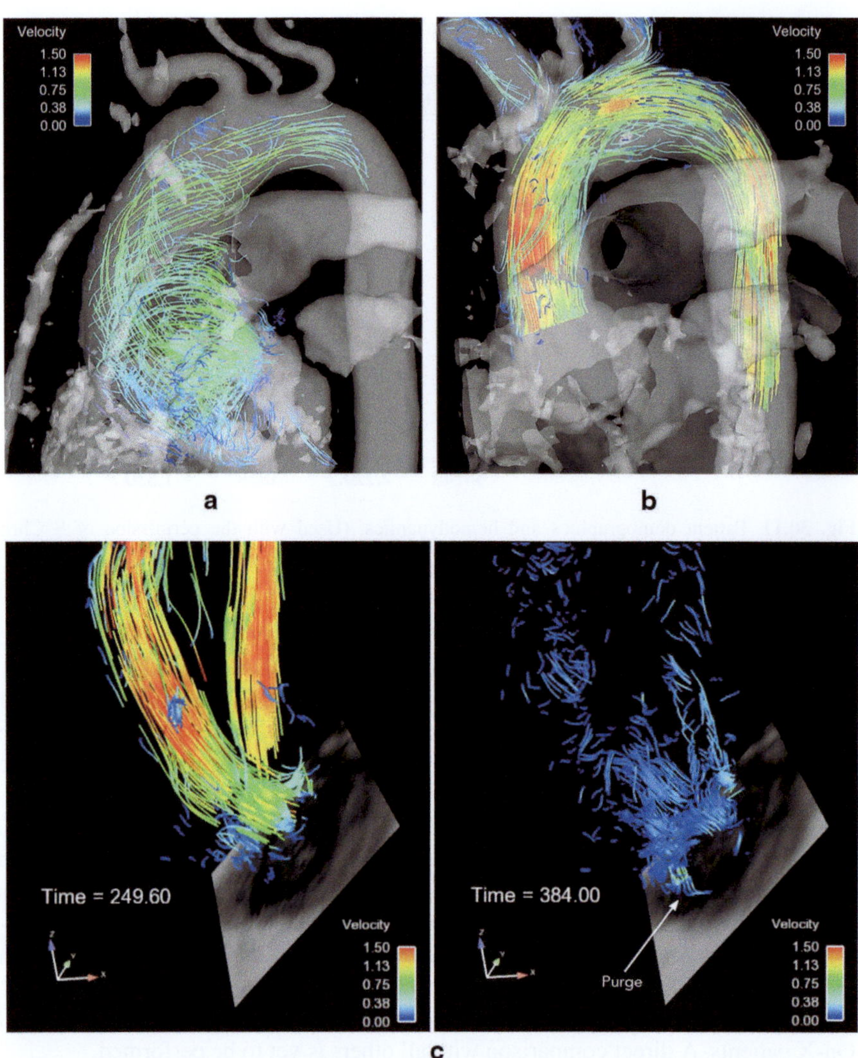

Fig. 30.12 (**a**) Pre-surgical magnetic resonance image. (**b**) Post-surgical magnetic resonance image with On-X valve replacement. (**c**) The streamline ejection offers a glimpse of the retrograde flow back through the tiny gap at the tips of the two leaflets. The pivot purge that provides the hydraulic cushion is visible at the start of diastole. (Used with the permission of S Chris Malaisrie MD)

forth in 1994 may have been realized and that the designers' objective to mimic natural flow with a carbon valve may have been attained.

The Berlin group's 4D-flow MRI characterizing turbulence to visualize in vivo blood flow was an important step in understanding turbulence and *normality*. It provides not only three-dimensional imaging but also the means of characterizing quantitatively the vorticity, helicity, and (in the ascending aorta) eccentricity. The latter is important in assuring the viability of the ascending aortic wall.

Although Philip Kilner had cautioned that magnetic resonance imaging detects only images that repeat beat by beat and does not reveal fine structure (such as small vortices or vortices within vortices or individual spinning cells), it is possible to compare *normality* with *abnormality* without knowing the details of the micro flow within the images. It's only necessary to match *normality* as revealed by 4D-flow MRI with the flow through a valve replacement using the same methodology.

The On-X valves' unprecedented hemodynamic performance protects *normal*, healthy flow of blood from disruption and allows *normal* spiral flow generated in the ventricle to enter the sinuses of Valsalva, as it does through the native valve. This process preserves momentum in order to produce vortical flow in the sinuses, as originally foreseen by da Vinci. The conserved rotational momentum ensures a healthy coronary flow reserve.

Hemodynamic precision doesn't happen by accident. Like many of da Vinci's revelations not fully appreciated during his lifetime, On-X carbon engineers and On-X valve designers have taken prosthetic heart valves beyond the current paradigm of having to choose between either flow or durability to give informed patients and their doctors more than has been thought possible in *one* valve. The resulting attributes include lifetime durability, the total biocompatibility and thromboresistance of pure carbon, *normal* spiraling flow with its vortices that conserve momentum to provide *normal* coronary flow reserve, protection from pannus encroachment, and FDA-approved lower anticoagulation requirement for the aortic valve.

The advent of the Berlin study had a profound impact on the chart, *Triumph Over Turbulence* (Fig. 30.13). The chart depicts a comparison of heart valve replacements used to replace deficient native valves:

Vertically:

1. The first vertical column labeled "Aortic Valve Type" lists the groups that are to be compared.
2. The second column labeled "Images" shows Doppler images on the left and MR images on the right.
3. The next set of four columns labeled "Flow *normality*" represent factors that are measurable and reflect the degree of *normality*:

   1. On the left are depicted the values of vorticity, helicity, or eccentricity. The values provide the degree of *normality* or *abnormality*. Values less than about 2 are usually *normal*. Values greater than about 2 represent departure from *normality*.
   2. The second column represents wall shear stresses, which are calculated from the values of vorticity, helicity, and eccentricity. High excessive wall shears can be pathologic.
   3. The third column is a measure of LDH, an enzyme that is released by red blood cells when damaged. For *normality*, LDH values must be in the *normal* range.
   4. The fourth column represents flow in the coronaries; for *normality*, it must be in the *normal* range.

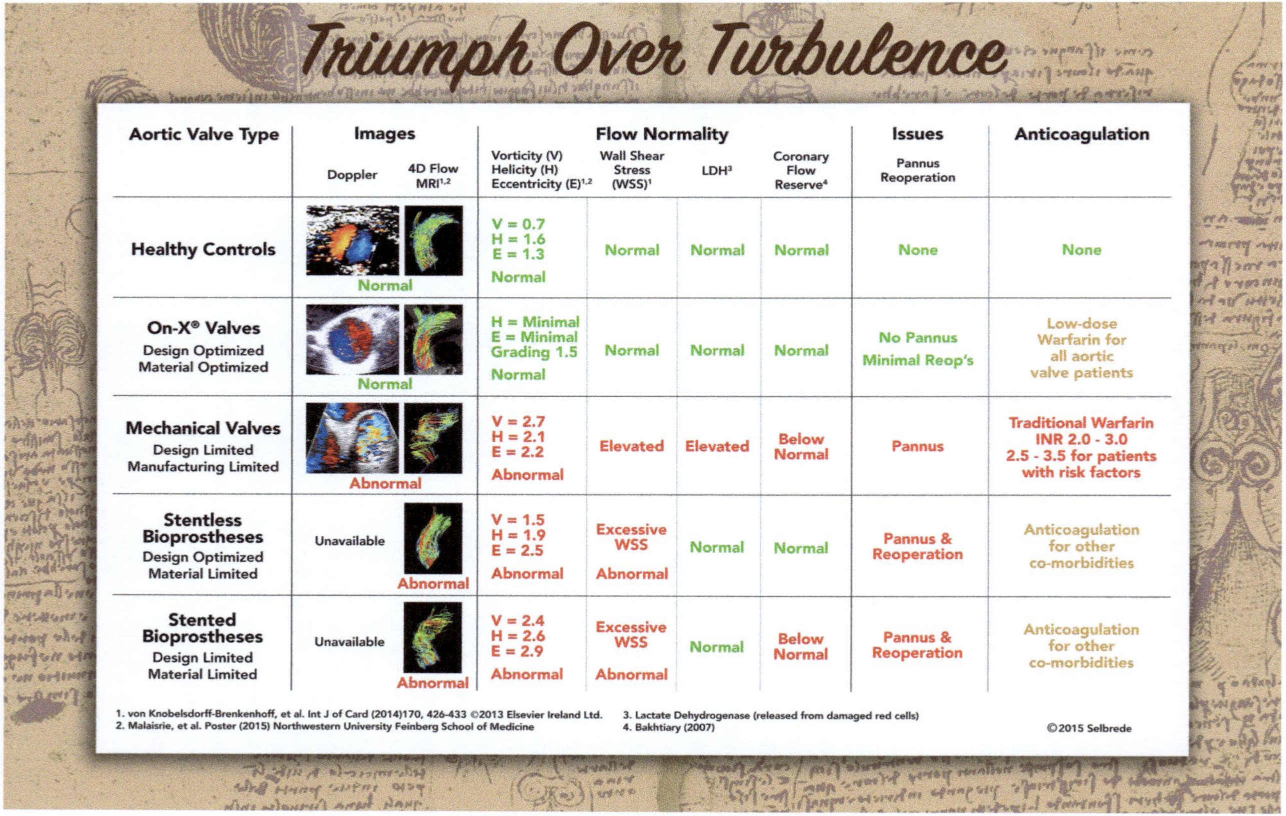

Triumph Over Turbulence

Aortic Valve Type	Images		Flow Normality				Issues	Anticoagulation
	Doppler	4D Flow MRI[1,2]	Vorticity (V) Helicity (H) Eccentricity (E)[1,2]	Wall Shear Stress (WSS)[1]	LDH[3]	Coronary Flow Reserve[4]	Pannus Reoperation	
Healthy Controls	Normal		V = 0.7 H = 1.6 E = 1.3 Normal	Normal	Normal	Normal	None	None
On-X® Valves Design Optimized Material Optimized	Normal		H = Minimal E = Minimal Grading 1.5 Normal	Normal	Normal	Normal	No Pannus Minimal Reop's	Low-dose Warfarin for all aortic valve patients
Mechanical Valves Design Limited Manufacturing Limited	Abnormal		V = 2.7 H = 2.1 E = 2.2 Abnormal	Elevated	Elevated	Below Normal	Pannus	Traditional Warfarin INR 2.0 - 3.0 2.5 - 3.5 for patients with risk factors
Stentless Bioprostheses Design Optimized Material Limited	Unavailable	Abnormal	V = 1.5 H = 1.9 E = 2.5 Abnormal	Excessive WSS Abnormal	Normal	Normal	Pannus & Reoperation	Anticoagulation for other co-morbidities
Stented Bioprostheses Design Limited Material Limited	Unavailable	Abnormal	V = 2.4 H = 2.6 E = 2.9 Abnormal	Excessive WSS Abnormal	Normal	Below Normal	Pannus & Reoperation	Anticoagulation for other co-morbidities

1. von Knobelsdorff-Brenkenhoff, et al. Int J of Card (2014)170, 426–433 ©2013 Elsevier Ireland Ltd.
2. Malaisrie, et al. Poster (2015) Northwestern University Feinberg School of Medicine
3. Lactate Dehydrogenase (released from damaged red cells)
4. Bakhtiary (2007)

©2015 Selbrede

Fig. 30.13 Performance comparison of aortic valve types and relative anticoagulation requirements (Used with the permission of Kathleen Selbrede)

4. The "Issues" column reports the presence or absence of pannus and/or reoperation.
5. The last column describes anticoagulation therapy.

Horizontally:

1. Horizontally, the control group is *normal* colored green.
2. For On-X valves, the items that match the control group are colored green. The last item colored yellow represents aortic patients that need a low dose of warfarin for non-cardiac risk factors.
3. The traditional mechanical valve data are consistently *abnormal* and colored red.
4. For the stentless group, both the LDH and coronary flow are *normal*. The V H E grade and wall shear stresses are *abnormal*.
5. For the stented group, only the LDH is *normal*. The V H E and wall shear stresses are *abnormal*.

The last items in both the bioprosthesis rows are split because some of these patients require anticoagulation for non-cardiac risk factors. The other patients, who receive no anticoagulation, are exposed to non-cardiac risk factors.

The Video and the In-House Critics

Just as we were finishing the video, the VC-CEO reminded me that he and his marketing team needed to review it for approval. By this time, the video was in final form: any changes would be expensive.

The word of a possible sale of On-XLTI to CryoLife increased the urgency of an accurate valuation. It wasn't long, however, before we learned that the sale price had already been "set in stone." So it was natural to conclude that the value to CryoLife would increase and the findings displayed in the video would enhance the marketability of On-X valves and the value of CryoLife.

The review meeting included the VC-CEO and several others that made up his marketing team.

The first 10 min of the new video included background and used a da Vinci theme. The design of the valve was presented step by step, following hemodynamic principles to achieve a design that sought to eliminate turbulence and preserve *normal* flow.

The next 10 min was a tutorial that demonstrated *normality* of blood flow in healthy patients, coming from the published works of Kilner and Markl.

By this point in the video, I could see the VC-CEO nervously twitching as he did when I had shown him the same data spread out on his conference table previously when he protested that he didn't get the relevance and terminated my informal presentation with "I don't have time" for this. He also reminded me that the *Triumph over Turbulence* chart might not merit his approval.

Fig. 30.14 The first heart valve replacement that achieves *normal* flow. (Used with the permission of S Chris Malaisrie MD)

When we got into the next sequence, starting with the Berlin study that involved 4D-flow MRI of representative heart valves, both mechanical and biological, the VC-CEO sat stiffly, looking straight at the imagery.

Anticipating that the new 4D-flow MRI revelations would not be immediately assimilated by those reviewing the video, we prepared a compilation of all the data into the summarizing chart, *Triumph Over Turbulence*, three words that echoed the achievement and ramifications set forth in the Berlin study and the poster from Northwestern University.

The video's conclusions substantiated that the plague of turbulence (absence of *normality*) that triggered the need for anticoagulation had been substantially removed by using On-X valves.

The video ends with a 4D-flow MRI movie depicting the preservation of *normal* flow through an On-X valve with a 20-s grand finale, accompanied by music composed by my son-in-law, Martin Selbrede, who among other things composed the orchestral background for the entire video (Fig. 30.14).

The VC-CEO was particularly distressed when the chart *Triumph over Turbulence* came on screen. I pointed out that a viewer lacking the appropriate technical background might need to be spoon-fed to achieve comprehension. Yet there was no passion to understand. The distressing responses from the VC-CEO and the marketing team to the revolutionary information set before them could perhaps be succinctly described as Buckberg had quipped regarding the surgeons he had tried to enlighten: "He looks, but does not see."

Comments coming from the reviewers were revealing:

1. Can't you dumb it down?
2. Perhaps some video clips might be effective.
3. How can we use it in marketing?

4. Doctors' attention span is only about 3 min!
5. It's really complicated!

Afterward, I sat the VC-CEO down and showed him the *Triumph over Turbulence* chart:

1. The first row colored green represents data that describes flow through the heart of healthy controls.
2. The second row represents data for On-X valves, likewise colored green, with the exception of the last item, colored yellow, that depicts a low dose of anticoagulation required for non-cardiac risk factors.
3. The third row, colored red, represents mechanical valves all with *abnormal* flow.

These three rows demonstrate that On-X valves should not be considered just another member of the mechanical replacement valve family. Instead, the On-X valve stands alone, alongside of the healthy controls, being the first valve replacement to uniquely enable the preservation of *normal* flow. I believe this monumental achievement was not properly recognized.

In fact, in the same week that I presented the video review, another 3-min video featuring the coach of the Chicago Bulls, who was lucky to have an On-X aortic valve replacement, was made public. All employees were invited to a lunch to view the Bulls' coach video, and the achievement of the valves' *Triumph over Turbulence* video was not mentioned.

In November 2015, I was invited to present our new video in Japan, at a meeting of the country's association for artificial organs. I was allowed 50 min for the video and discussion. The auditorium had 150 seats. Over half were occupied by Japanese investigators of artificial hearts and cardiac assist devices, among the world's most intellectually astute experts in blood flow through devices and healthy human hearts.

The video triggered a standing ovation, with one person in the question period simply rising long enough to assert, "I am going to use your valve."

In the spring of 2016, I received the publication [6] that contained all the information that appeared in the Malaisrie Bio poster from Northwestern, Fig. 30.1. The only significant difference was that the published version did not include the isolated aortic replacement: patient numbers 7 and 10 were not included, and only patients with aortic root replacement were included.

Their conclusions: "Preliminary evidence suggests the aortic root replacement with the On-X mechanical valve significantly reduces aberrant aortic hemodynamics, producing patterns that resemble those in healthy volunteers."

The meaning was simple. In spite of the aortic root replacement that included a conduit, the On-X valve produced the same *normality* as was depicted in patients with isolated aortic valve replacement that was furnished by the Malaisrie poster in 2015.

Shortly after the video completion, I submitted to the VC-CEO an expense report for my payments to the vendor. He refused to pay and passed it over to CryoLife.

Also, Kathy and Martin, who were not paid for their efforts in producing the video including Martin's background music, presented to the VC-CEO a statement of their copyrights. He refused to sign. His signature however was not required as they were not paid for anything and their rights remain.

After the sale was closed, I presented my expenses for the production of the video to CryoLife and was reimbursed.

References

1. Malaisrie SC, Barker, et al. Restoration of physiologic flow in the ascending aorta after mechanical aortic valve replacement. In: Poster from Northwestern University Feinberg School of Medicine presented at proceedings of the heart valve society. Monaco; 2015.
2. Von Knobelsdorff-Brenkenhoff F, et al. Blood flow characteristics in the ascending aorta after aortic valve replacement—a pilot study using 4D-flow MRI. Int J Cardiol. 2014;170(3):426–33.
3. Kilner PJ, Yang GZ, Mohiaddin RH, et al. Helical and retrograde secondary flow patterns in the aortic arch studied by three-directional magnetic resonance velocity mapping. Circulation. 1993;88:2235–47.
4. Kilner PJ, Yang GZ, Wilkes AJ, Mohladdin RH, Firmin DN, Yacoub MH. Asymmetric redirection of flow through the heart. Nature. 2000;404(6779):759–61.
5. Puskas J, Gerdisch M, Nichols D. Reduced anticoagulation after mechanical aortic valve replacement: interim results from the prospective randomized On-X valve anticoagulation clinical trial randomized Food and Drug Administration investigational device exemption trial. J Thorac Cardiovasc Surg. 2014;147:1202–10.
6. Keller EJ, Malaisrie SC, Krose J, McCarthy PM, Carr JC, Markl M, Barker AJ, Collins JD. Reduction of aberrant aortic haemodynamics following aortic root replacement with a mechanical valve conduit. Interact Cardio Vas Thorac Surg. 2016;23:416–23. https://doi.org/10.1093/icvts/ivw173.

Chapter 31
CryoLife Inc. Buys On-X Life Technologies Inc.

On January 20, 2016, CryoLife's purchase of On-XLTI closed. I was offered a consultancy that would continue for a year.

J. Bokros, *Heart of Carbon*, https://doi.org/10.1007/978-3-031-17933-4_31

Chapter 32
Mervyn Williams's South African Trial of the On-X Valve

In January 2017, I received Williams's manuscript that was presented at the annual meeting of the Asian Society for Cardiovascular and Thoracic Surgery, Seoul, South Korea, March 2017 [1]. Williams's manuscript, with his permission, is verbatim.

The On-X Heart Valve at 15 Years in a Poorly Anticoagulated Population.
M. A. Williams, S. Van Riet, and M. Jason
Used with permission of Mervyn Williams

Introduction

The population of South Africa is both culturally and ethnically complex. Historical imbalances have produced a population that is largely socio-economically disadvantaged. Poor nutrition and overcrowding together with a high unemployment rate (30–50%) and substandard education have resulted in a population where infectious diseases such as rheumatic fever are common. To these realities a high incidence of tuberculosis and HIV/AIDS add further problems.

Shortly after the establishment of the cardiac unit in Port Elizabeth in 1982 it became apparent that the management of anti-coagulation in patients requiring mechanical heart valve replacement was problematic. Overdosing with Warfarin was recognized as a major concern resulting in a high incidence of bleeding with a number of anticoagulant-related deaths. Our response was to lower the level of anti-coagulation, aiming at an International Normalised Ratio (INR) of between 1.5 and 2.5. Bleeding became much less of a problem, but we still recorded thromboembolism, and valve thrombosis was a frequent reason for emergency surgery.

When the On-X valve became available in the late 90s we believed that its design and the use of a newly developed pyrolytic carbon (On-X carbon) would mitigate the incidence of thromboembolic complications and valve thrombosis. On-X carbon is silicon free and elimination of silicon from the manufacturing process results

J. Bokros, *Heart of Carbon*, https://doi.org/10.1007/978-3-031-17933-4_32

in a much smoother and theoretically less thrombogenic surface. The leaflet housing is of a tubular configuration rather than the washer configuration used in other devices. The valve has a flared inlet and a natural length to diameter ratio. This design, when combined with fully opening leaflets, results in organized flow with minimal turbulence. In the aortic position it preserves the normal spiral flow from the left ventricle to the aorta. Our early experience comparing the On-X valve with the Medtronic, Inc., Hall and CarboMedics, Inc. valves confirmed a lower incidence of adverse events and the On-X valve became, and is currently, our valve of choice [2].

Between October, 1999 and November, 2014 a total of 1387 valves were implanted in 1187 patients. Among these patients 346 had aortic valves replaced, 641 had mitral valves replaced and 189 had both mitral and aortic valves replaced. The patient demographics are listed in Table 32.1. The etiology of the valve disease is detailed in Table 32.2, the most common cause being rheumatic carditis.

Surgical Procedure

All valves were implanted using similar techniques. Surgery was undertaken using standard cardiopulmonary bypass with moderate hypothermia and crystalloid cardioplegic arrest.

Table 32.1 Patients' clinical characteristics

Patients' clinical characteristics			
	AVR	MVR	DVR
No of Patients	346	641	189
Age (years)	45 (9–87)	38 (3–84)	31 (9–80)
Gender (% female)	37	59	50
Follow up (pt-yrs)	1618	3205	885
Mean	4.4	5.0	4.4
Sinus Rhythm	73%	56%	59%

Table 32.2 Etiology (%)

Etiology (%)			
	AVR	MVR	DVR
Rheumatic	29%	71%	77%
Infective endocarditis	11%	7%	10.5%
Degenerative	41%	1.6%	5%
Previous valve	8.6%	10%	7%
Congenital	7.5%	0.5%	0.8%

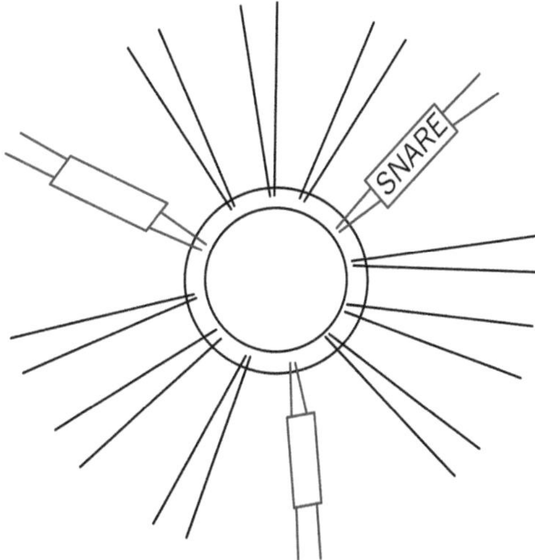

Fig. 32.1 Seat the aortic valve by placing snares on the three sutures at the nadir of the sinuses and seating the valve by tightening the valve by tightening the snares

In the mitral position the valve was implanted using everting pledgeted sutures. In the aortic position the valve was inserted either with simple interrupted sutures or with vertical pledgeted mattress sutures. The valve was oriented in the anti-anatomic position in the mitral annulus, whilst in the aortic position the valve was placed perpendicular to the ventricular septum. Many surgeons (ourselves included) encountered technical difficulties when first implanting the aortic valve. This is usually due to oversizing of the valve. It should be remembered that the effective orifice of the On-X valve is the same as most valves two sizes larger.

We have found it useful to seat the aortic valve by placing snares on the three sutures at the nadir of the aortic sinuses and seating the valve by tightening the snares (Fig. 32.1). This maneuver seats the valve within the annulus and the other sutures can be tied easily, leaving the 3 snared sutures till last. In double valve implantations we have by preference used the 25–33 (one size fits all) valve in the mitral position. The placing of the sewing cuff close to the outlet orifice of the valve places the valve in the atrium without projection into the left ventricular outflow tract, leaving the outflow tract free of any potential obstruction.

The unreliable nature of the study population made follow up difficult and somewhat erratic. These difficulties have been compounded in recent years by a marked deterioration in the state medical services and facilities. An ailing economy combined with a generally corrupt and inefficient administration has significantly diminished the money available for health services.

Most clinics are dysfunctional, record keeping has become chaotic and in many cases, it was difficult and often not possible to obtain patient records. Fortunately, almost everybody in South Africa now has a mobile phone and much of the data

Table 32.3 Patients' anticoagulation status

Patients' anticoagulation status	
Attended regularly, INR 1.5–2.5	47%
Attended irregularly, INR outside range	25%
No anticoagulants or Unknown status	28%

Table 32.4 Causes of valve-related deaths

Causes of valve related deaths					
	Endocarditis	CVA	Bleeding	Thrombosis	Unknown
MVR	6	9	2	5	21
AVR	4	7	1	0	10
DVR	4	9	0	1	7
Total	14	25	3	6	38

from rural patients was obtained telephonically. All patients seen or contacted during 2014 were included for the purpose of statistical analysis of late results. Despite these challenges follow up was 75% complete.

Before discharge, all patients were anti-coagulated using Warfarin, the target INR being 1.5–2.5. Patients who attended regularly at clinics and whose INR values fell within this range were regarded as adequately anti-coagulated. Most of the patients from rural areas did not attend clinics and were not anti-coagulated. The anticoagulation status of the patients is detailed in Table 32.3.

Operative events were summarized as simple percentages equal to the number of events divided by the number of patients. Valve-related events were defined according to the published guidelines for reporting mortality and morbidity after valve operations [3]. Results were expressed as linearised rates obtained by dividing the number of late events by the total patient years of follow up. Actuarial event free rates were compiled using the Kaplan-Meier product limit method [4].

Results

Overall hospital mortality was 1.7%. None of the deaths was valve-related.

Among 171 deaths which occurred during the follow up period 89 were valve-related (Table 32.4). Included in the total of 89 deaths are 53 deaths where the cause of death is not known. Overall actuarial survival at 15 years is shown in Fig. 32.2 and Kaplan Meier prediction of freedom from valve-related death at 15 years is in Fig. 32.3.

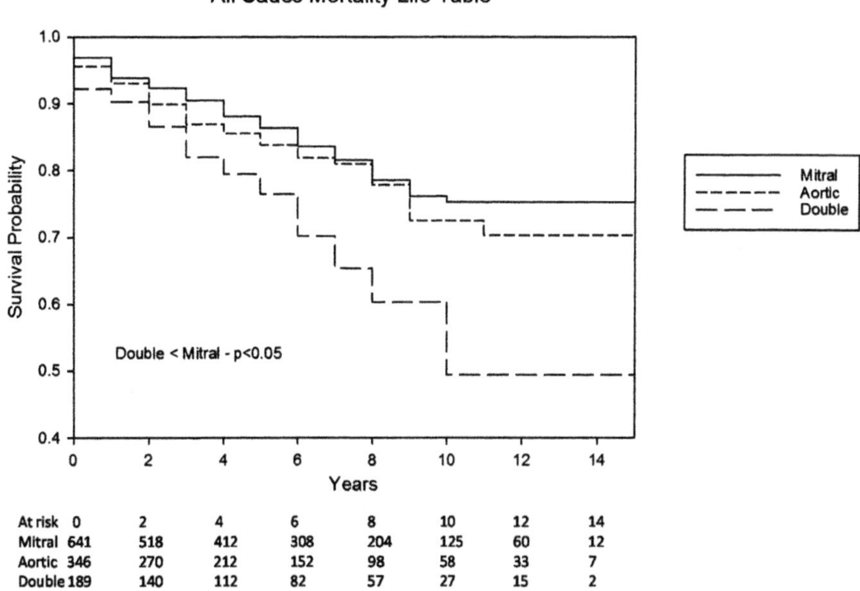

Fig. 32.2 Overall actuarial survival at 15 years

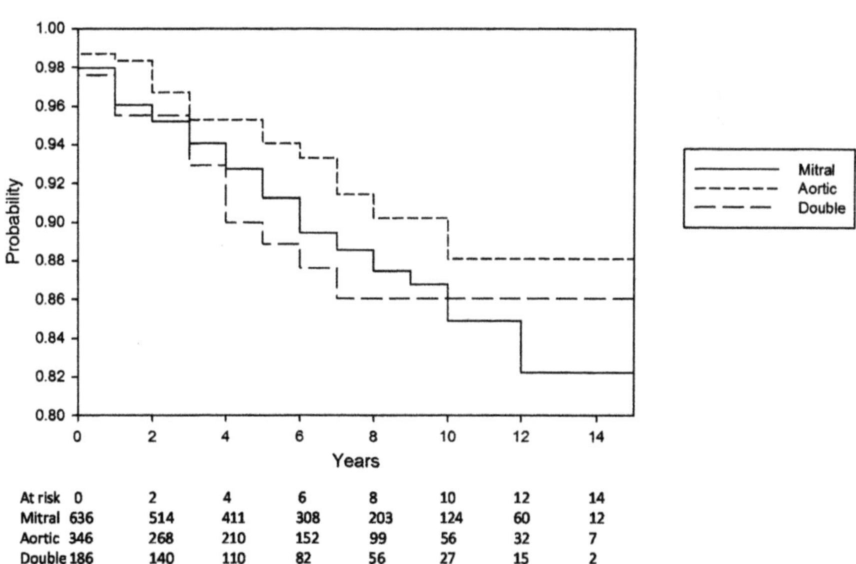

Fig. 32.3 Actuarial freedom from valve-related death at 15 years

Table 32.5 Adverse event rates

Adverse event rates	AVR	MVR	DVR
Thromboembolism			
Total	0.87(14)	1.18(39)	1.58(14)
Minor/recovered	0.43(7)	0.56(18)	0.45(4)
Major	0.43(7)	0.66(21)	1.13(10)
Thrombosis	0	0.22(7)	0.11(1)
Bleeding event	0.06(1)	0.12(4)	0.45(4)
Endocarditis	0.25(4)	0.28(9)	0.45(4)
Paravalvular leak	0.12(2)	0.06(2)	0
Explant	0.06(1)	0.12(4)	0.23(2)
Valve related death	1.17(19)	1.68(54)	1.8(16)
Hospital deaths	2%(7)	0.9%(6)	1.1%(2)

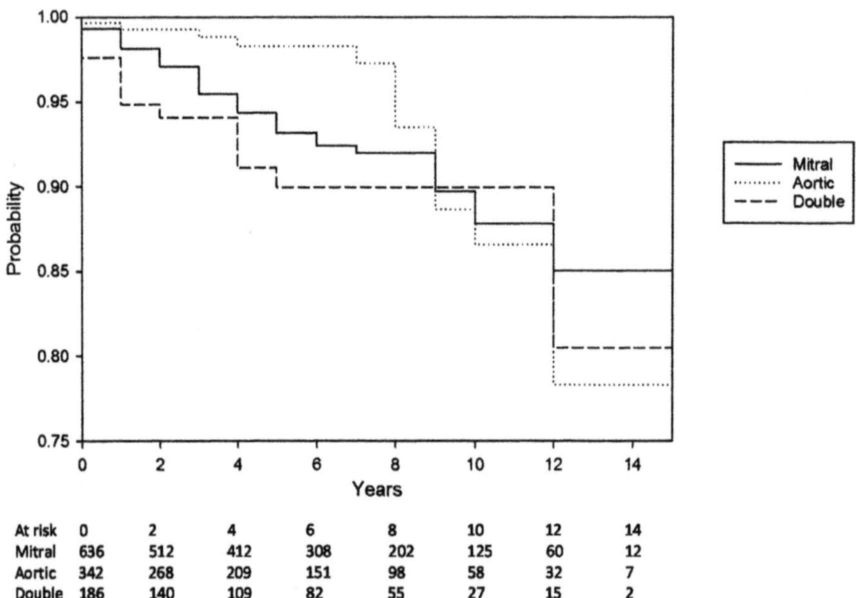

Fig. 32.4 Actuarial freedom from thromboembolism at 15 years

Thromboembolism There were 67 systemic embolic events. 37 were major events resulting in death or severe residual disability. The linearised rates for thromboembolism are shown in Table 32.5 and the actuarial prediction of freedom from thromboembolism at 15 years in Fig. 32.4.

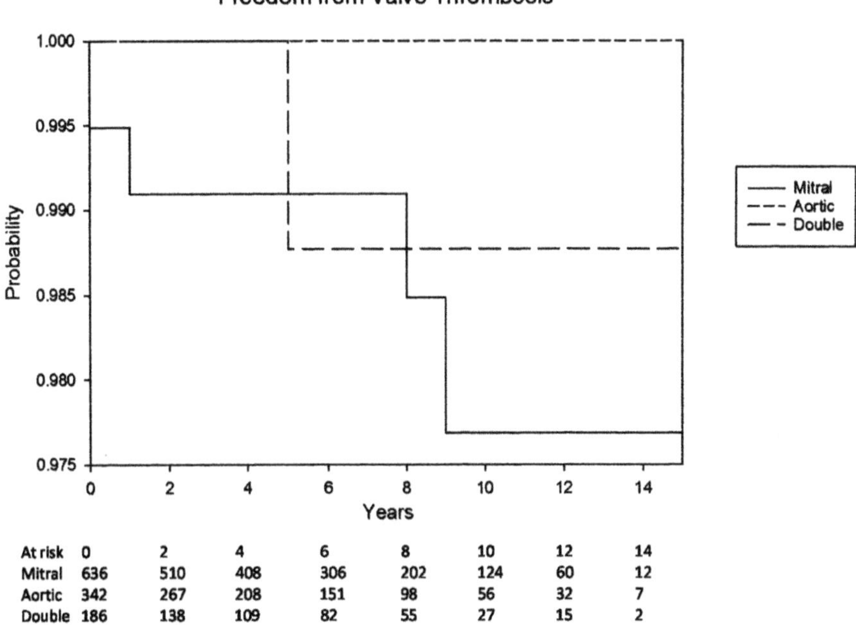

At risk	0	2	4	6	8	10	12	14
Mitral	636	510	408	306	202	124	60	12
Aortic	342	267	208	151	98	56	32	7
Double	186	138	109	82	55	27	15	2

Fig. 32.5 Actuarial freedom from valve thrombosis at 15 years

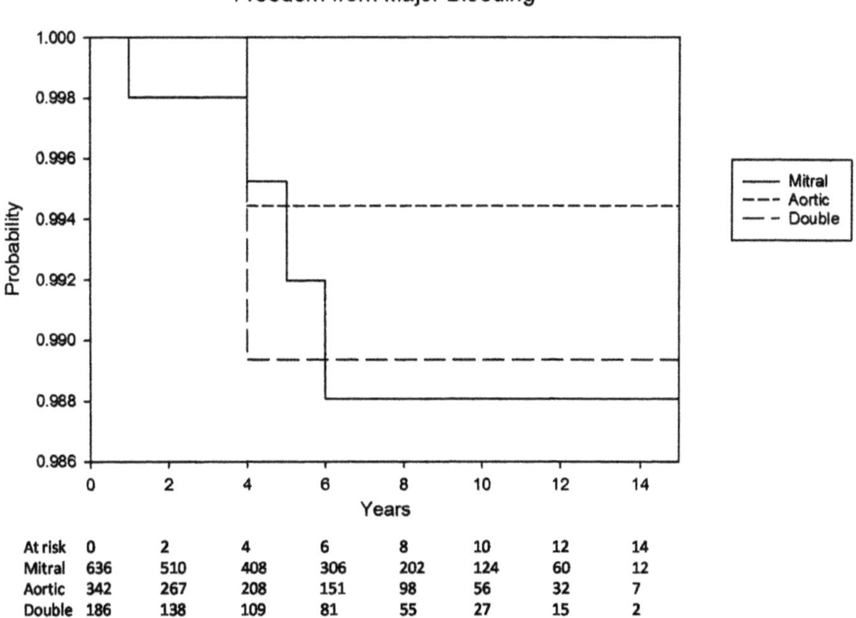

At risk	0	2	4	6	8	10	12	14
Mitral	636	510	408	306	202	124	60	12
Aortic	342	267	208	151	98	56	32	7
Double	186	138	109	81	55	27	15	2

Fig. 32.6 Actuarial freedom from bleeding at 15 years

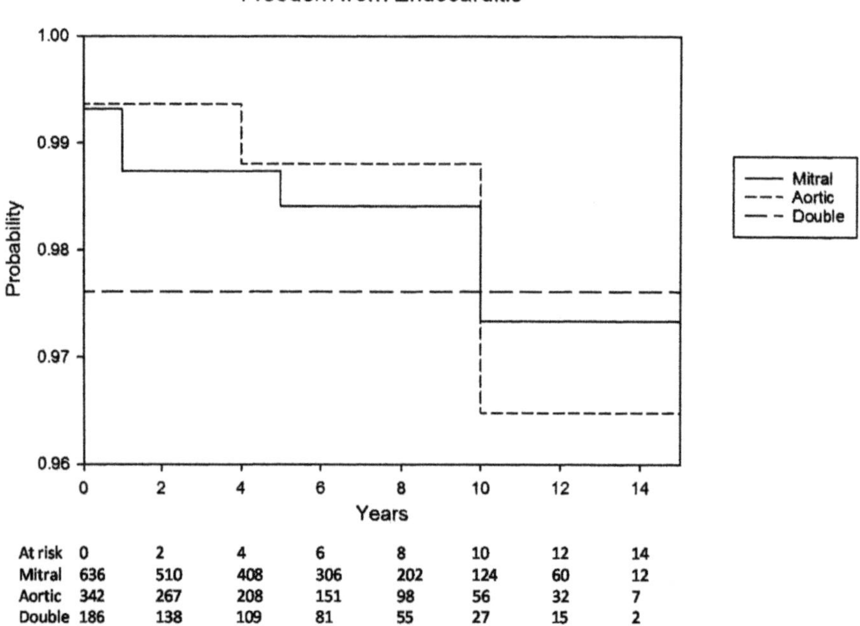

At risk	0	2	4	6	8	10	12	14
Mitral	636	510	408	306	202	124	60	12
Aortic	342	267	208	151	98	56	32	7
Double	186	138	109	81	55	27	15	2

Fig. 32.7 Actuarial freedom from endocarditis at 15 years

Thrombosis There were seven thrombosed valves and three of these patients died. One patient underwent re-replacement with an On-X valve at 1 year. The second valve thrombosed 6 years later resulting in her death. Two mitral valves thrombosed shortly after the delivery of a full-term infant. Both of these patients died. In both cases the anti-coagulation following delivery was mismanaged. The linearised rate for valve thrombosis is shown in Table 32.5 and the actuarial freedom from thrombosis in Fig. 32.5.

Bleeding Events Minor bleeds such as nose bleeds and bleeding gums were not regarded as valve-related and are not recorded. There were six major bleeding events three of which were fatal. Linearised rates for bleeding are shown in Table 32.5 and the actuarial freedom from bleeding in Fig. 32.6.

Endocarditis Endocarditis was a common reason for valve surgery (see Table 32.2). Fourteen patients died from infective endocarditis, most of them from ongoing disease. Actuarial freedom from endocarditis at 15 years is shown in Fig. 32.7.

Pregnancy There were 67 full term pregnancies. Two patients died after delivery. These deaths were related to thrombosis of the valve. The post-delivery management of anti-coagulation was mismanaged in both cases.

Discussion

This is Williams's third report on the use of the On-X valve in a poorly anti-coagulated, largely developing nations group. Previous reports were at 5 and 10 years [5, 6]. It is the largest group so far reported on the On-X valve—1187 patients followed for 5708 pt years. Follow up extends to 15 years with a mean follow up of 4.4–5 years. This report is limited by the difficulties in late surveillance. In our report at 5 years follow up was 95% and at 10 years was 85%. It is currently 75%. There are many reasons for the progressive drop in patient follow up. Many attend hospital only when they are ill, travel is expensive, and impoverished unemployed patients simply cannot afford to go to a clinic. These factors are compounded by a failing public health service with dysfunctional clinics and hospitals, often unable to supply medication.

Notwithstanding, our findings cannot be ignored—the large number of patients and years of follow up are compelling factors. In addition the rates of adverse events and valve-related deaths are similar to the rates recorded in our previous reports, suggesting that those patients lost to follow up would have had outcomes similar to those followed for up to 15 years.

Valve thrombosis is a catastrophic event with a significant mortality rate. In well anticoagulated patients this complication is uncommon. There is evidence that some ethnic groups are more at risk of this complication than others. For example, Reddy reported a high incidence of valve thrombosis in Indian patients when the CarboMedics, Inc. valve was used [7]. In contrast no thrombosed valves were recorded from Japan where the same valve was used and the level of anticoagulation was of low intensity [8, 9]. We and others have recorded high rates of valve thrombosis in inadequately anticoagulated South African patients when other valves were used [2, 10, 11].

Use of a prosthetic valve in a population of developing nations delineates performance differences that may not be apparent in First World populations. Small differences in thrombotic events are exaggerated, unmasking the true thrombogenic potential of different valves. Although our patients are anticoagulated at a lower level than recommended by most valve companies, the incidence of major adverse prosthesis-related events is similar to the incidence reported from First World communities [12–15]. Our low incidence of valve thrombosis (0.22% patient year in the mitral position and 0.00 in the aortic position) has made a welcome change from our previous experience with other valves.

Predictably, the lower dose of anticoagulants used in our patients reduces significantly the incidence of bleeding complications. Our data together with other published series prompted the Prospective Randomised On-X Anticoagulation Clinical Trial (PROACT) to test the safety of lower dose anticoagulation. This trial documented a 65% reduction in bleeding without a significant increase in thromboembolism [16]. The authors cautioned that the results should not be extrapolated to other prostheses.

We confirmed in a previous publication that other mechanical prostheses had a higher rate of thrombotic complications than the On-X valve when lower dose anticoagulation was used [2]. A number of publications have documented low adverse event rates, low transvalvular gradients, and minimal hemolysis using the On-X valve [17–21]. Chan and Jamieson found a significant improvement in mitral and aortic valve morbid event rates compared with other mechanical valves [22]. Chambers recorded low adverse event rates in a large series (691 patients) followed for a maximum of 12.6 years [23].

The On-X valve is currently approved by the Federal Drug Administration (FDA) for use in the aortic position with low dose anticoagulation (INR 1.5–2). The low incidence of bleeding complications suggests that, with lower dose anticoagulation, the On-X valve can be used safely in the aortic position at all ages including the elderly.

Valve obstruction by pannus has not been encountered. None of the published reports on the On-X valve have recorded this complication and On-X Life Technologies are not aware of a single valve obstruction by pannus. Pannus obstruction has been recorded in all of the other currently available mechanical and biological heart valves [24–28]. The absence of pannus obstruction is related to valve design. The valve housing is of a tubular configuration as compared to the washer-like configuration of other mechanical valves. This design feature has serendipitously prevented the obstruction of the valve orifice by pannus. *(As an Addendum, more recently CryoLife identified two cases with one of them associated with off-label use of an On-X valve implanted in the tricuspid position.)*

The first percutaneous Transcatheter Aortic Valve Implantation (TAVI) in a human was reported by Cribier et al in December 2002 [29]. Grube made a similar claim in December 2005 [30] and the initial experience with transapical implantation was reported in 2006 [31].

TAVI was carried out in patients where the risk of surgery was prohibitive. When confined to this group of patients TAVI showed a significant improvement in short-term survival and notwithstanding a higher rate of stroke and other complications and was accepted as the proper management of aortic stenosis in inoperable patients.

Following the initial reports there have been well over 2000 publications related to TAVI. The introduction of TAVI rekindled enthusiasm for bioprostheses in patients younger than 65 years—spurred by the proposal that structural valve deterioration (SVD) could be managed by "valve in valve" TAVI.

It is well documented that SVD is age-related and the younger the patient the greater its rapidity [32–34]. In children and young adults the outcomes can be catastrophic [35]. The Mayo Clinic showed that patients aged 50–70 years who underwent aortic valve replacement with mechanical valves had a survival advantage relative to matched patients who received bioprostheses [36]. Others have reported similar findings [37, 38]. Prior to the advent of TAVI it was generally accepted that bioprostheses should be used only in patients older than 65–70 years. Results relating to a cohort of patients with an average age of 83 years managed by TAVI was presented recently. A significant increase in degeneration was observed between 5 and 7 years after TAVI and it was estimated that it would be 50% at 8 years [39]. If TAVIs deteriorate at this rate in 80-year-olds what can we expect in younger patients?

All bioprostheses deteriorate with age and considering the published experience we believe it remains unwise to use bioprosthetic valves including TAVI in patients younger than 65–70 years. It is illogical to believe that TAVIs will behave differently than other biological valves.

The reason for using bioprostheses in the elderly is the perceived danger of anticoagulants. This danger has been exaggerated and fueled by tissue valve proponents referring to Coumadin as "rat poison."

Sharabiani, et al. reported their experience with mechanical valves in septuagenarians [40]. They observed that because of increasing life expectancy elderly patients with bioprostheses were requiring re-operation for SVD. They elected to use mechanical valves in the elderly and concluded there was no increased risk of in-hospital mortality, bleeding, or thromboembolic events. They noted their concern that TAVI is being extended to low-risk and intermediate risk surgical patients where no compelling evidence for its efficacy exists.

Vicchio, reporting his experience using mechanical and biological valves in octogenarians, noted only one anticoagulant bleed in 98 patients receiving mechanical valves [41]. Actuarial survival at 8 years was significantly better in the mechanical group. The PROACT study has shown that in the aortic position the On-X valve can be used safely with low intensity anticoagulation (INR 1.5–2) and should be considered as an alternative to bioprostheses in the aortic position in elderly patients.

There have been 67 full-term pregnancies. Two deaths following delivery were related to mismanagement of anticoagulation therapy in the post delivery period. We believe that with the proper management of anticoagulation during pregnancy the On-X valve can be used safely in women of childbearing age.

Conclusion

Bearing in mind the nature of the population with poor anti-coagulation coverage these results are regarded as excellent. They confirm our earlier reports suggesting that the On-X valve should be the valve of choice when anti-coagulation coverage is uncertain. Apart from recording low rates of valve-related complications with INR levels of 1.5–2.5 a major advantage is the complete absence of pannus obstruction, a feature clearly related to valve design, and unique to the On-X valve. We believe that the On-X valve can be used safely in the aortic position in elderly patients.

References

1. Williams MA, Van Riet S, Jason M. The On-X heart valve at 15 years in a poorly anticoagulated population. In: Presented at the annual meeting of the Asian Society for Cardiovascular and Thoracic Surgery in Seoul, South Korea. 2017.
2. Williams MA, Crause L, Van Riet S. A comparison of mechanical valve performance in a poorly anticoagulated community. J Card Surg. 2004;19:410–4.
3. Clark RE, Edmunds LH Jr, Cohn LH, et al. Guidelines for reporting morbidity and mortality after cardiac valvular operations. Eur J Cardiothorac Surg. 1988;2:293.
4. Kaplan EL, Meier P. Non parametric estimation from incomplete observations. J Am Stat Assoc. 1958;53:457–81.
5. Williams MA, Van Riet S. The On-X heart valve mid term results in a poorly anticoagulated population. J Heart Valve Dis. 2006;15:80–6.
6. Williams MA, Van Riet S. The On-X Valve at 10 years. In: Presented at the 6th Biennial Meeting of the Society for Heart Valve Disease June 25–28 2011, Barcelona.
7. Reddy N, Padmanabhan T, Singh S, et al. Thrombolysis in left-sided prosthetic valve occlusion immediate and follow up results. Ann Thorac Surg. 1994;58:462–71.
8. Soga Y, Okabayashi H, Nishina T. Up to 8-year follow-up of valve replacement with CarboMedics, Inc. valve. Ann Thorac Surg. 2002;73:474–9.
9. Kurisu K, Tominaga R, Ochiai Y, et al. A 10-year experience with the CarboMedics, Inc. cardiac prosthesis. Ann Thorac Surg. 2005;79:784–9.
10. Antunes MJ, Wessels A, Sadowski RG, et al. Medtronic, Inc. Hall valve replacement in a third world population group. J Thorac Cardiovasc Surg. 1988;95:980–93.
11. Deviri E, Sareli P, Wisenbaugh T, Cronje SL. Obstruction of mechanical heart prostheses clinical aspects and surgical management. J Am Coll Cardiol. 1991;17:646–50.
12. Emery RW, Krogh CC, Arom KV, et al. The St. Jude Medical cardiac valve prosthesis: a 25-year experience with single valve replacement. Ann Thorac Surg. 2005;79:776–82.
13. Butchart EG, Li HH, Payne N, Buchan K, Grunkenmeir GL. Twenty years experience with the Medtronic, Inc. Hall valve. J Thorac Cardiovasc Surg. 2001;121:1090–100.
14. Toole JM, Stroud MR, Kratz JM. Twenty-five year experience with the St. Jude medical mechanical valve prosthesis. Ann Thorac Surg. 2010;89:1402–9.
15. Van Nooten GJ, Caes F, Francois K, et al. Twenty years single-center experience with mechanical heart valves: a critical review of anticoagulation policy. J Heart Valve Dis. 2012;21:88–98.
16. Puskas J, Gerdisch M, Nichols D. Reduced anticoagulation after mechanical aortic valve replacement: interim results from the prospective randomized On-X valve anticoagulation clinical trial randomized Food and Drug Administration Investigational Device Exemption trial. J Thorac Cardiovasc Surg. 2014;147:1202–10.
17. Laczkovics A, Heidt M, Oelert H, et al. Early experience with the On-X prosthetic heart valve. J Heart Valve Dis. 2001;10:94–9.
18. Moidl R, Simon P, Wolner E. The On-X prosthetic heart valve at five years. Ann Thorac Surg. 2002;74:S1312–7.
19. Palatianos G, Laczkovics A, Simon P, et al. Multicentered European study on the safety and effectiveness of the On-X prosthetic heart valve: intermediate follow-up. Ann Thorac Surg. 2007;83:40–6.
20. McNichols KW, Ivey T, Metras J, et al. North American multicenter experience with the On-X prosthetic heart valve. J Heart Valve Dis. 2006;15:73–9.
21. Tossios P, Reber D, Oustria M, et al. Single centre experience with the On-X prosthetic heart valve between 1996 and 2005. J Heart Valve Dis. 2007;16:551–7.
22. Chan V, Jamieson WRE, et al. Influence of the On-X mechanical prosthesis on intermediate term major thromboembolism and hemorrhage A prospective multicentre study. J Thorac Cardiovasc Surg. 2010;140:1053.

23. Chambers JB, Pomar JL, Mestres CA, Palatianos GM. Clinical event rates with the On-X bileaflet mechanical heart valve: a multicentre experience with follow-up to 12 years. J Thorac Cardiovasc Surg. 2013;145:420–4.
24. Imasaka K, Oe M, Baba H, Sumida H. Obstruction of the left ventricular outflow tract due to pannus formation after the implantation of a CarboMedics, Inc. aortic valve. Jpn J Thorac Cardiovasc Surg. 2006;54:304–7.
25. Teshima H, Aoyagi S, Hayashida N. Dysfunction of an ATS valve in the aortic position: the first reported case caused by pannus formation. J Artif Organs. 2005;8:270–3.
26. Teshima H, Fukunaga S, Takaseya T. Obstruction of St. Jude medical valves in the aortic position: plasma transforming growth factor type beta 1 in patients with pannus overgrowth. Artif Organs. 2010;34:210–5.
27. Naito Y, Hachida M, Shimabukuro T. St. Jude Medical prosthetic aortic valve malfunction due to pannus formation. Jpn J Thorac Cardiovasc Surg. 2000;48:739–41.
28. Hirota M, Isomura T, Yoshida M. Subvalvular pannus overgrowth after mosaic bioprosthesis implantation in the aortic position. Ann Thorac Cardiovasc Surg. 2016;22:108–11.
29. Cribier A, Eltchaninoff H, Basharal A. Percutaneous transcatheter implantation of an aortic valve prosthesis for calcific aortic stenosis: first human case description. Circulation. 2002;106:3006–8.
30. Grube E, Laborde JC, Zickmann B, et al. First report on a human percutaneous transluminal implantation of a self-expanding valve prosthesis for interventional treatment of aortic valve stenosis. Catheter Cardiovasc Interv. 2005;66:465–9.
31. Lichtenstein SV, Cheung A, Ye J, et al. Transapical aortic valve implantation in humans: initial clinical experience. Circulation. 2006;114(6):591–6.
32. Une D, Ruel M, David TE. Twenty-year durability of the aortic Hancock II bioprosthesis in young patients: is it durable enough? Eur J Cardiothorac Surg. 2014;46:825–30.
33. David TE, Feindel CM, Bos J, et al. Aortic valve replacement with Toronto SPV bioprosthesis: optimal patient survival but suboptimal valve durability. J Thorac Cardiovasc Surg. 2008;135:19–24.
34. Daneshmand MA, Milano CA, Rankin JS. Influence of patient age on procedural selection in mitral valve surgery. Ann Thorac Surg. 2010;90:1479–85.
35. Williams MA. Tissue valves in young patients – a recipe for disaster. J Card Surg. 1991;6:620–3.
36. Brown ML, Schaff HV, Lahr BD, et al. Aortic valve replacement in patients aged 50 to 70 years: improved outcome with mechanical versus biologic prostheses. J Thorac Cardiovasc Surg. 2001;358:878–84.
37. Kulik A, Bedard P, Lam BK, et al. Mechanical versus bioprosthetic valve replacement in middle-aged patients. Eur J Cardiothorac Surg. 2006;30:485–91.
38. Weber A, Noureddine H, Englberger L, et al. Ten-year comparison of pericardial tissue valves versus mechanical prostheses for aortic valve replacement in patients younger than 60 years of age. J Thorac Cardiovasc Surg. 2012:1075–83.
39. Dvir D. First look at long-term durability of transcatheter heart valves: assessment of valve function up to 10 years after implantation Presented at the European Association of Percutaneous Cardiovascular Interventions (EuroPCR) Paris 17–20 May 2016.
40. Sharabiani MT, Fiorentino F, Angelini GD, Patel NN. Long-term survival after surgical aortic valve replacement among patients over 65 years of age. Open Heart. 2016;3(1):e000338.
41. Vicchio M, Della Corte A, DeSanto LS, et al. Tissue versus mechanical prostheses: quality of life in octogenarians. Ann Thorac Surg. 2008;85:1290–5.

Chapter 33
Summary and Reflections

The history of carbon heart valve replacements was all but rewritten with the arrival of the Williams report.

I first met Mervyn Williams in London in 1999 and introduced our valve to him. It piqued his interest sufficiently to provoke him and his wife Bernadette to initiate their pivotal study of the valve in South Africa.

At first, South Africa's non-compliant patient population made follow-up a challenge. But 15 years of data gathering with a highly non-compliant population revealed a hitherto unseen truth. Although the Williamses were unable to control their patients' anticoagulation therapy, those patients still did as well as strictly controlled valve patients from first world populations. This well-attested result was prima facie evidence that On-X valves were actually preserving native valve blood flow: a truly revolutionary result.

In light of the 15 years of data he amassed, Williams was confident in drawing two far-reaching conclusions. First, he proposed that women wanting children should consider the On-X valve as a preferable option. Second, he states that for the elderly, the risk-to-benefit ratio for TAVI versus a mechanical valve favors the On-X valve.

The photo in Fig. 33.1, taken in 2012 at the Lahey Clinic in Boston, shows three major contributors to the development of heart valve replacements standing alongside me. Vincent Gott (a presenter at the event) stands between Sherry Shaw-Smith and me while holding his children's book illustrating various hand-drawn surgical procedures. Vince does not lack for career options in his retirement.

Sherry was deeply involved with cardiovascular start-up companies for four decades, among them Abiomed, Edwards, Marquette Electronics, and Argon. When she joined On-X Life Technologies in 2004, she distinguished herself by promoting

Supplementary Information The online version contains supplementary material available at https://doi.org/10.1007/978-3-031-17933-4_33. The videos can be accessed by scanning the related images with the SN More Media App.

J. Bokros, *Heart of Carbon*, https://doi.org/10.1007/978-3-031-17933-4_33

Fig. 33.1 Sherry Shaw-Smith; Vincent Gott MD; Jack Bokros; and Prof. Ned Hwang. (Used with the permission of Sherry Shaw-Smith)

the On-X valve with integrity and passion. By 2010, both Sherry and Vince were insisting that the carbon story was a tale that needed to be told in a definitive way by someone at ground zero. Their appeals were persuasive. I reluctantly agreed to put my hand to that task.

Also in Fig. 33.1, at the far right, is Professor Ned HC Hwang, an expert in fluid mechanics specializing in blood flow. Ned was a major contributor to the development of the On-X valve. His earliest directive to our team set an important but simple precedent: "Don't fight the flow; follow the flow." Gerry Buckberg, noted authority on the form and function of the heart who assisted us with design, told me, "I like your Chinese friend. His *don't fight, just follow* ... profoundly simple, but right on." Buckberg's assessment proved insightful. An example is in Fig. 5.5 where the white ball is silicone rubber which is denser than blood; the black ball's density matches that of blood and produces less turbulence.

Although the low-dose anticoagulation trial was initiated in 2006, approval was not granted until 2015. As that trial plodded along at a snail's pace, focus shifted to the video (Chap. 28) that my wife, Roberta, and daughter, Kathy Selbrede, with her husband, Martin (Fig. 33.2), had set out to produce. Buckberg, learning the video was a family project, chipped in by introducing us to Philip Kilner of the Yacoub group (the Royal Brompton) and Markl of the Northwestern University, noted experts in 4D MRI visualization of blood flow. The video (Fig. 33.3), an independent endeavor, was applauded and warmly received across the globe by everyone.

I'll conclude by sharing the two single most important events of the foregoing story.

Fig. 33.2 Martin Selbrede and Kathleen Selbrede. (Used with the permission of Jack Bokros)

The first occurred in 1992 during the Medtronic Project (Chap. 18). At that time, the control of the carbon process was an impediment restricting valve design. David Wilde approached me with an idea that might overcome the existing fabrication obstacles and the design limits they imposed. I urged him to pursue his idea. Within a month, David and Jim Accuntius, a colleague, presented us with a graph (Fig. 18.9). Its impact was momentous.

The principle driving this new innovation reminded me of a story my grandfather[1] told me when I was 5 years old (Fig. 33.4). He was a lumberjack that tended the horses pulling timber to the river for transport to the sawmill. His story concerned legendary lumberjack Paul Bunyan and his blue ox, Babe. Bunyan faced a massive logjam caused by one misaligned log holding up all the rest. Once Bunyan identified the obstructing log, he attached a line to it so his ox could extract it. The impediment now cleared, the logs thundered down the river to float serenely into the sawmill.

Just as Bunyan had broken that logjam wide open, David Wilde's idea opened the door to permit the fabrication of the On-X valve by turning our contraption into a precision research device.

The second event, and the one with the most impact on our development, was the Medtronic clinical trial (Chap. 18). After all the pre-clinical testing was successfully completed, a clinical trial of the Parallel Valve was carried out to evaluate its viability in human patients. We expected that the trial would be designed to confirm safety and effectiveness using a protocol that was approvable by the FDA.

In any event, we persevered to produce the most important advance in the history of heart valve replacement technology. We're looking forward to the future, a future where ethical, academic, and professional integrity will mark the evaluation of revolutionary advances in medical technology.

---

[1] My grampa used to hoist me onto his back for a bumpy horse trot ride. I called him "Bumpy." I loved Bumpy.

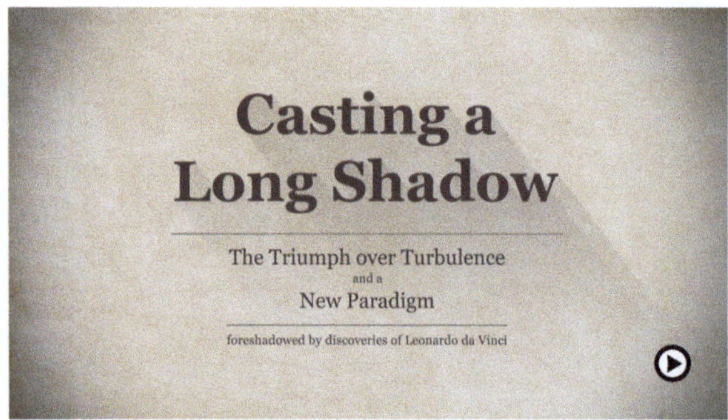

Fig. 33.3 Video "Casting a Long Shadow" (▶ https://doi.org/10.1007/000-033)

Fig. 33.4 Grampa "Bumpy" Anderson and Jack Bokros. (Used with the permission of Jack Bokros)

Appendix

APPENDIX

Used with permission of Ling Ma

Studies on Pyrolitic Carbon

For Biomedical Applications

PhD Dissertation by

Ling Ma

UCLA

UNIVERSITY OF CALIFORNIA

Los Angeles

Studies on Pyrolytic Carbons

for Biomedical Applications

A dissertation submitted in partial satisfaction of the

requirements for the degree Doctor of Philosophy

in Materials Science and Engineering

by

Ling Ma

CHAPTER
TWELVE

Summary and Conclusions

As the entirety of this dissertation gradually materialized, the research project that lasted more than six years has finally come to an end. In this chapter finale, it seems appropriate to restate the highlights of this project that was often associated with the number of cycles that appeared eternally endless. These highlights are also summarized graphically in Fig. 33.1 at the end of this chapter.

Reliability of Pyrolytic Carbon for Heart Valve Application

For the first time, such a large number of pyrolytic carbon specimens have been tested to ultra-high cycles; 146 specimens were tested to a total number of cycles that exceeded 9.7×10^{10}. This is equivalent to 2630 years if one cycle is one heart beat.

Because of this laborious effort, questions that had lingered over the reliability of pyrolytic carbon can now be answered. As a federal regulatory agency, the Food and Drug Administration (FDA) issued specific requirements on pyrolytic carbon for heart valves applications. It stated:

> …statistically valid data must be presented to determine the fatigue strength of the material at 6×10^8 cycles. This fatigue strength must be defined by a minimum of 90% survival with 95% confidence. Furthermore, it must be shown that a minimum safety factor of 2 exists between the peak service stress and the fatigue strength at 6×10^8 cycles.

Here are the results obtained from this research:

1. all Si-alloyed pyrolytic carbon specimens survived more than 6×10^8 cycles at stress levels that were at least two times the peak service stress. The survival probability was 92.2% with 95% confidence.
2. all unalloyed pyrolytic carbon specimens survived more than 6×10^8 cycles at a stress level that was four times the peak service stress. The survival probability was 91.6% with 95% confidence.
3. furthermore, a group of Si-alloyed specimens survived more than 6×10^8 cycles having sizable cracks ranging from 0.60 to 1.84 mm induced in each specimen. All these specimens were loaded at two times the peak service stress. No crack growth or specimen failure occurred, thereby giving 90.5% survival probability with 95% confidence.
4. also, a group of unalloyed specimens survived more than 6×10^8 cycles with a crack (ranging from 0.866 to 1.581 mm) induced in each specimen. All were loaded at two times the peak service stress. No crack growth or specimen failure occurred, thus giving a 90.5% survival probability with 95% confidence.
5. finally, in spot tests performed at two times and four times the peak service stress, specimens survived more than 2.2×10^9 cycles without failure.

Therefore, the pyrolytic carbon materials tested in this study not only satisfied the FDA requirement, but also survived more severe conditions: four times the peak stress and sizable cracks.

However, a 90% survival probability for a heart valve is hardly impressive, especially when one in every ten patients receiving heart valve implantation surgery is at risk of losing their life. The acceptable risk for such implantation surgery should be less than one in a million. To achieve such a high probability means to test millions of specimens with each specimen to a billion cycles. This is practically impossible.

On the other hand, the key point in interpreting the above test data is the test stress level. By testing specimens at a higher stress level than the service level with an attainable probability, the results would imply an ultra-high survival probability for pyrolytic carbon in heart valve applications at the service stress.

Fatigue Crack Growth Threshold

Also for the first time, it was demonstrated that there is a threshold value of stress intensity factor below which a crack in the isotropic pyrolytic carbon would not grow for ultra-high cycles. The significance of this demonstration is that the threshold was obtained experimentally rather than by extrapolation of crack growth rate. The lowest fatigue crack growth rate ever published was around 10^{-10} m/cycle. For heart valve applications, a growth rate at or below 10^{-14} m/cycle is desired. To use the extrapolation shortcut, one needs to extrapolate data by four to five orders of magnitude. Due to the uncompromising nature of the application, it is paramount to collect data experimentally whenever possible instead of by extrapolation.

The existence of this threshold is very significant in quality control and reliability of the heart valves made of pyrolytic carbons. When the flaw size is less than the threshold size the flaw would not propagate during the service lifetime, provided that the valve is operated under the designed stress. The threshold size of fatigue crack non-propagation makes the proof test of the components and the whole assembly for the valve very effective to screen out the flaws because proof tests can be performed at a level one order of magnitude higher than the service stress.

Does Pyrolytic Carbon Fatigue At All?

This question comes up naturally after the impressive results of the ultra-high cycles fatigue test. To answer this question, groups of specimens were tested at extremely high cyclic stress levels that were within the scatter band of the fracture strength of the materials. Two distributions were compared, one from the static strength test, the other from the strength after millions of cycles at high stress levels. The latter distribution fell within the static strength distribution, indicating that the millions of cycles at extremely high stress level did not degrade the strength of the material.

Spot tests were also conducted at the maximum cyclic stress level that is equal to 90–95% of the mean static strength. Specimens accumulated more than a billion cycles without failure.

These all support the notion that pyrolytic carbon is insensitive to cyclic stressing all the way up to its static strength. The results also suggest that the cyclic stress at such a high level did not cause small flaws and voids to initiate cracks. They further indicate that the existing micro cracks in the material did not propagate even under this extremely high cyclic stress.

Probability of Failure and Weibull Statistics

The results of stepped fatigue tests in Chap. 8 show that the fatigue strength distribution and static strength distribution of this isotropic pyrolytic carbon are basically the same. The fatigue strength distribution was analyzed using Weibull statistics and two important conclusions were drawn:

1. by extrapolating the strength distribution down to the peak service level of the heart valve, the cumulative probability of failure at 250 $\mu\varepsilon$ was $1-10^{-0.000000000000000001259}$. This is an infinitesimal number and can be regarded as zero in practice. In other words, to operate mechanical heart valves made of this isotropic pyrolytic carbon at a peak service level of 250 $\mu\varepsilon$, the probability of fatigue failure due to intrinsic flaws in the material is almost nil (approaching zero).
2. as illustrated in Table 8.7, the cumulative probability of failure decreased by three orders of magnitude when the stress level was lowered by a factor of two. This relationship quantitatively validated the approach of testing a reasonable manageable number of samples at higher stress levels to ensure ultra-high survival probability at the stress level of interest.

Micro Cracks

Concerns over micro cracks were outlined in Chaps. 3 and 9. The core issue was that micro cracks in some ceramic materials would propagate under a range of stress intensity factor ΔK that was two to three times smaller than the ΔK_{th}. Is this also the case for micro cracks in pyrolytic carbon? Here were the findings:

1. a fatigue crack growth threshold was observed from fatigue-grown micro cracks. Crack sizes ranged from 136 to 170 μm. The threshold value ΔK_{th} for micro cracks is comparable to the values for larger cracks.
2. the critical stress intensive factor K_{1c} estimated from specimens with micro cracks yielded results of 1.76 MPa$\sqrt{m}$. This is close to 1.67 MPa$\sqrt{m}$ measured by using compact disk specimens with a much larger crack in the same material.
3. fatigue tests of specimens with micro cracks (ranging from 13.8 to 22.2 μm) at extremely high stress levels (within the scatter band of fracture strength) showed that failures were not initiated from these cracks. The resulting fracture strain after the cyclic stressing also points out that these cracks did not degrade the materials' strength.
4. in a spot test, a specimen with a crack of 13.3 μm was tested at the maximum cyclic strain of 8000 $\mu\varepsilon$. The applied ΔK was 1.08 MPa$\sqrt{m}$, 85% of the critical stress intensity factor K_{1c} measured from larger cracks. The specimen accumulated 9.15×10^8 cycles without crack growth.

In summary, the micro cracks in pyrolytic carbons, alloyed or unalloyed, do not behave differently from their larger counterparts. This finding is very important due to the fact that microscopic voids and flaws are unavoidable in CVD of pyrolytic carbon and the fact that microscopic voids and flaws are difficult to detect.

Microstructure of Pyrolytic Carbon

The microstructural analysis indicated that the turbostratic feature in pyrolytic carbon occurred on a very fine scale. The average size of pyrolytic carbon "crystallites" is about 3–5 nm. It is difficult to imagine dislocations playing any significant role in influencing mechanical properties. If there are indeed dislocations in pyrolytic carbon, as other researcher have argued, they would run into barriers every 3 or 5 nm; thus, making the dislocations immobile. On the other hand, because turbostratic pyrolytic carbon is highly disordered, one might say there are dislocations everywhere if the dislocation is defined as an imperfection in a crystal structure. In the latter case, the moving of dislocations (if they are mobile) is not significant since wherever they go there are already plenty of dislocations. So far, no evidence has shown that by just moving the dislocations one can turn turbostratic pyrolytic carbon into diamond or graphite.

In any event, the lack of crack initiation mechanisms (normally associated with mobile dislocations) might be the reason why the pyrolytic carbon is insensitive to cyclic stressing all the way up to its fracture strength.

Conclusions

The isotropic pyrolytic carbons studied in this research project are an excellent material for artificial heart valve applications. They possess very favorable fatigue (or non-fatigue) behavior that ensures a heart valve would function for decades. Their fracture mechanics properties are normal and predictable when compared to other types of ceramic materials. So conventional methods, such as the proof test, can be used in controlling the quality and reliability of the product. However, mechanical heart valve designers are cautioned to consider the effects of extrinsically introduced flaws. Mishaps have occurred when unfavorable hydrodynamic conditions caused cavitation in the blood to damage the valve surface, thereby leading to the fracture of leaflets in the implanted heart valves

It is worthwhile to point out that the pyrolytic carbon is one of the pioneering ceramic-like materials used as structural components in critical applications. The achievements are impressive with over three million pyrolytic components used in mechanical heart valves, and an accumulated experience well in excess of 15 million patient years. While the designers are perfecting the pyrolytic carbon in mechanical heart valve applications, other possible uses of this unique material should be explored (Fig. 33.A.1).

Summary of Studies on Pyrolytic Carbons

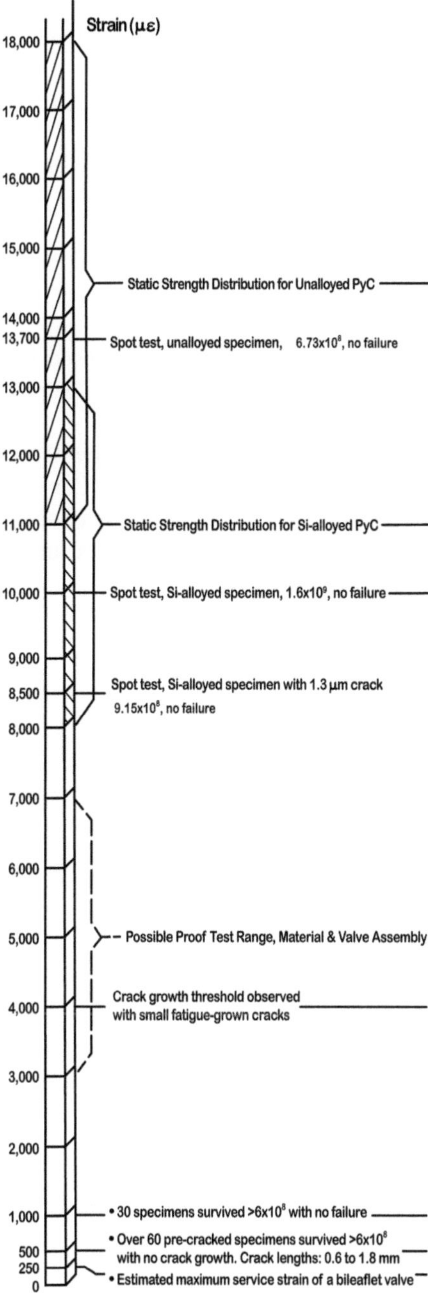

Fig. 33.A.1 Summary of studies on pyrolytic carbons. (Used with the permission of Ling Ma)

Publications emanating from Ma's thesis

1. Sines G and Ma L. Long-life cycle fatigue of pyrolytic carbon in Bioceramics: Proceedings of the 6th International Symposium on Ceramics in Medicine, Philadelphia, Nov 1993;211–15
2. Ma L and Sines G. Threshold size for cyclic fatigue crack propagation in a pyrolytic carbon Materials Letters 1993;17:49–53
3. Ma L, Sines G, Gilpin B. Threshold crack size in a pyrolytic carbon for no-growth under cyclic stress Materials Research Society Proceedings 1, Diamond and other forms of carbon 1995;383:273–9
4. Ma L, Sines G and Gilpin B. On fatigue and fracture behavior of Si-alloyed pyrolytic carbon, Proceedings of 6th International Symposium on the Fracture of Ceramics Plenum Press 1996;12:107–120
5. Ma L and Sines G. Fatigue of isotropic carbon used in implanted mechanical heart valves Special issue of J Heart Valve Dis. Supplement 1 June 1996;5: S59–S65
6. Ma L and Sines G. Fatigue growth threshold in pyrolytic carbon. Observations from small fatigue-grown cracks, 23rd Biennial Conference on Carbon 18–23 July 1997—Extended Abstracts, American Carbon Society, 562–3
7. Ma L and Sines G. Unalloyed pyrolytic carbon for implanted mechanical heart valves, J Heart Valve Dis. 1999;8(5):578–85
8. Ma L and Sines G. Fracture behavior of an unalloyed pyrolytic carbon. J of Biomed. Mater Res. 2000;51:61–8
9. Ma L and Sines G. High resolution structural studies of pyrolytic carbon used in medical applications Carbon 2002;40:445–54

References

Bokros JC, Rosenthal PC. Liquid emulsion autoradiography of metals. J Metals. 1956;8:286.

Bokros JC. Critical recrystallization of zirconium. Trans AIME. 1960;218:351.

Bokros JC. Effect of sodium on the mechanical properties of zirconium. Corrosion. 1961;17:131.

Bokros JC, Wallace WP. High temperature, high pressure oxidation of alloys caused by carbon dioxide. Corrosion. 1960;16:117.

Bokros JC. Graphite-metal compatibility at high temperatures. J Nucl Mater. 1961;3:89.

Meyer RA, Bokros JC. Graphite-metal and graphite-molten salt systems. Chpt 15. In: Nightengale RE, editor. Graphite. New York: U.S. Atomic Energy Commission, Academic; 1962. p. 445.

Bokros JC. Creep properties of uranium-zirconium-hydrogen alloy. J Nucl Mater. 1961;3:216.

Bokros JC. Transformation kinetics of a zirconium-uranium-hydrogen alloy. J Nucl Mater. 1961;3:320.

Bokros JC, Merten U. Hydrogen migration in zirconium-hydrogen-uranium alloys in thermal gradients. J Nucl Mater. 1963;10:201.

Bokros JC, Parker ER. Mechanism of the martensite burst transformation in Fe-Ni single crystals. Acta Metall. 1963;11:1291.

Oxley JH, Secrest AC, Veigel ND, Blocher JM. J Am Inst Chem Engrs. 1961;7(3):498.

Franklin RE. The interpretation of diffuse x-ray diagrams of carbon. Acta Cryst. 1950;3(2):107–21.

Franklin RE. On the structure of carbon. J de Chimie Physique et de Physico-Chimie Biologique. 1950;47(5,6):573–5.

Franklin RE. A rapid approximate method for correcting the low-angle scattering measurements for the influence of the finite height of the x-ray beam. Acta Cryst. 1950;3(2):158–9.

Franklin RE. Influence of the bonding electrons on the scattering of x-rays by carbon. Nature. 1950;165(4185):71–2.

Bokros JC. The structure of pyrolytic carbon deposited in a fluidized bed. Carbon. 1965;3:17.

Bokros JC. Absorption factors for a modified Bacon preferred-orientation technique. Carbon. 1965;3:167.

Bokros JC. Variations in the crystallinity of carbons deposited in fluidized beds. Carbon. 1965;3:201.

Bokros JC. Deposition, structure, and properties of pyrolytic carbon, Chapter 1. In: Walker PL, editor. Chemistry and physics of carbon. New York: Marcel Dekker, Inc; 1969. p. 5.

Bokros JC. Random pyrolytic carbon. Nature. 1964;202:1004.

Goeddel WV, Bokros JC. High temperature nuclear fuels. In: Holden AN, editor. AIME nuclear metallurgy symposium. Delavan, Wisconsin, October 1966, vol. 42, 1966. New York: Gordon and Breach. p. 85–104.

Gott VL, Koepke DE, Daggett RL, et al. The coating of intravascular plastic prostheses with colloidal graphite. Surgery. 1961;50(2):382–9.

Gott VL, Daggett RL, Koepke DE, et al. Replacement of canine pulmonary valve and pulmonary artery with a graphite coated valve prosthesis. J Thorac Cardiovasc Surg. 1962;44(6):713–21.

Gott VL, Whiffen JD, Dutton RD, et al. The anticlot properties of graphite coatings on artificial heart valves. In Abstract of paper submitted for presentation at the Sixth American Carbon Conference, Pittsburgh, PA, June 17–21, 1963.

Gott VL, Whiffen JD, Dutton RD. Heparin bonding on colloidal graphite surfaces. Science. 1963;142:1297–8.

Hastings FW, Harmison LT. Proceedings of artificial heart program (Hegyeli RJ, editors). National Heart Institute, June 9–13, 1969.

Bokros JC, Ellis WH. US Patent No 3,526,005, Carbon medical implant devices, issued September 1, 1970.

De Laszlo. US Patent No 3,526,906, issued September 8, 1970.

Bokros, JC. US Patent No 3,579,645, Broad medical implants, issued May 25, 1971.

Bokros, JC. US Patent No 3,676,179, Residual stress, issued July 11, 1972.

Bokros JC, Ellis WH. US Patent No 3,677,795, PyC Medical devices, issued July 18, 1972.

Bokros JC, Ellis WH. US Patent No 3,685,059, Medical devices, issued August 22, 1972.

Bokros JC, Ellis WH. US Patent No 3,707,006, Orthopedic implant devices, issued December 26, 1972.

Bokros JC. US Patent No 3,783,868, Percutaneous device, issued January 8, 1974.

Bokros JC. US Patent No 3,971,134, Dental implant, issued July 27, 1976.

Bokros JC. US Patent No 4,131,957, Finger joint, issued January 2, 1979.

Bokros JC, Haubold AD, Akins RJ, et al. Trends in prosthetic heart valve design. In: Bodnar E, Frater RWM, editors. Replacement cardiac valves. New York: Pergamon Press, Inc; 1991. p. 333–55.

Bokros JC, Haubold AD, Akins RJ, et al. The durability of mechanical heart valve replacements: past experiences and current trends. In: Bodnar E, Frater RWM, editors. Replacement cardiac valves. New York: Pergamon Press, Inc.; 1991. p. 21–48.

Haubold AD, Bokros JC. Carbon in medical devices, Chapter 1. In: Williams DF, editor. Biocompatibility of clinical implant materials, CRC series in biocompatibility, vol. 2. Boca Raton: CRC Press; 1981.

Bokros JC, Gott VL, LaGrange LD, et al. Heparin sorptivity and blood compatibility of carbon surfaces. J Biomed Mater Res. 1970;4:145–87.

LaGrange LD, Gott VL, Bokros JC, et al. Compatibility of carbon and blood. In: Hegyeli RJ editors. Proceedings of artificial heart program. National Heart Institute, Washington, DC, June 9–13, 1969.

Bokros JC, Gott VL, LaGrange LD, et al. Correlations between the blood compatibility and heparin sorptivity for an impermeable, isotropic pyrolytic carbon. J Biomed Mater Res. 1969;3:497–528.

Gott VL, Alejo DE, Cameron DEMD. Mechanical heart valves: 50 years of evolution. Ann Thorac Surg. 2003;76:S2230.

Gott VL, Daggett RL, Young WP. Development of a carbon coated central-hinging bileaflet valve. Ann Thorac Surg. 1989;48:528–30.

Bokros JC. US Patent No 3,546,711, Heart Valve, issued December 15, 1970.

Zlotnick AY, Shiran A, Lewis BS, Aravol D. A perfectly functioning Magovern-Cromie sutureless prosthetic aortic valve 42 years after implantation. Circulation. 2008;117(1):e1–2.

Bard RJ, Baxman HR, Berting JP, et al. Carbon. 1968;6:603.

Bokros JC, Akins RJ. US Patent No 3,977,896, Steady State Bed, issued August 31, 1976.

Bokros JC, LaGrange LD, Schoen FJ. Control of carbon structures. In: Walker Jr PL, Thrower PA, editors. Bioengineering, chemistry and physics of carbon, vol. 9. New York: Marcel Dekker Inc.; 1973. p. 103–71.

Brewer III LA, editors. Proceedings of second national conference on prosthetic heart valves. May 30–31, June 1, 1968 (editors). Springfield: Charles Thomas; 1969.

Scott SM, Sethi GK, Paulson DM, Takaro T. Insidious strut fractures in a DeBakey-Surgitool aortic valve prosthesis. Ann Thorac Surg. 1978;25(4):382–4.

Von de Emden J, Eberiein U, Breme J. Asymptomatic strut fracture in DeBakey-Surgitool aortic valves Texas. Heart J. 1990;17(3):223–7.

Beiras-Fernandez A, Oberhuffer M, Kur F, et al. 34-year durability of a DeBakey-Surgitool mechanical aortic valve prosthesis. Interact Cardiovasc Thorac Surg. 2006;5:637–9.

Butany J, Naseemuddin A, Feindel CM. DeBakey-Surgitool mechanical heart valve prosthesis, explanted at 32 years. Cardiovasc Pathol. 2004;13(6):345–6.

Beall AC Jr, Bloodwell RD, Liotta D, et al. Elimination of the sewing ring-metal seat interface in mitral valve prostheses. Circulation. 1968;37(Suppl II):184.

Beall AC Jr, Bloodwell RD, Arbegast NR, et al. Mitral valve replacement with Dacron-covered disc prosthesis to prevent thromboembolism: clinical experience in 202 cases, Chapter 21. In: Brewer III LA, editor. Prosthetic heart valves. Springfield: Thomas; 1970.

Beall AC Jr, Morris GC Jr, Noon GP, et al. An improved mitral valve prosthesis. Ann Thorac Surg. 1973;15(1):25–34.

Beall AC Jr, Morris GC Jr, Howell JF Jr, et al. Clinical experience with an improved mitral valve prosthesis. Ann Thorac Surg. 1973;15(6):601–6.

Fernandez J, Chang K, Gooch A, et al. Anatomic and clinical analysis of 96 Beall prostheses explanted over a 13 year period. Chest. 1983;83:632–7.

Topaz O, Rutherford MS, Mackey-Bojack S. Beware of the B(e)all valve. Tex Heart Inst J. 2010;37(2):237–9.

Merendino AK, editors. Proceedings of conference on prosthetic heart valves for cardiac surgery, September 9–10, 1960. Springfield: Charles Thomas; 1961.

IR-100 Award, Special carbon for artificial heart valves. Industrial Inc. 1971.

Pyrolite, serial no 312,673, filed Nov 20, 1968, Carbon coated prosthetic devices, heart valves and pumps, vascular tubes and hip joint balls, class 44 (int. Cl. 10) First use Aug 6, 1968, in commerce Aug 6, 1968.

Bokros JC, Guthrie GL, Schwartz AS. The influence of crystallite size on the dimensional changes induced in carbonaceous materials by high temperature radiation. Carbon. 1968;6:55.

Bokros JC, Guthrie GL, Dunlap RW, Schwartz AS. Radiation-induced dimensional changes and creep in carbonaceous materials. J Nucl Mater. 1969;31:25–47.

Bokros JC, Dunlap RW, Schwartz AS. Effect of high neutron exposure on the dimensions of pyrolytic carbons. Carbon. 1969;7:143.

Bokros JC, Koyama K. Interpretation of dimensional changes caused in pyrolytic carbon by high-fluence neutron irradiation. J Appl Phys. 1970;4(5):2146.

Letter to Circulation 76;53:2016

Starek PJK. Advantages and disadvantages of mechanical valves, Chapter 18. In: Starek, editor. Heart valve replacement and reconstruction: clinical issues and trends. Chicago: Year Book Medical Publishers, Inc; 1986. p. 221–35.

Callaghan JC, Coles J, Damie A. Six year clinical study of the use of the Omniscience valve prosthesis in 219 patients. J Am Coll Cardiol. 1987;9(1):240–6.

Teijeira FJ. Long-term experience with the Omniscience cardiac valve. J Heart Valve Dis. 1998;7(5):540–7.

Misawa Y, et al. Twenty-two year experience with the Omniscience prosthetic heart valve. ASAIO J. 2004;50:606–10.

Kazui T, Yamada O, Yamagisi M. Aortic valve replacement with Omniscience and Omnicarbon valves. Ann Thorac Surg. 1991;52:236–44.

Misawa Y, Hasegawa T, Kato M. J Thorac Cardiovasc Surg. 1993;150(1):168–72.

Weiss P, Jenzen HR, Hoffmann A, et al. The Omnicarbon tilting-disc heart valve prosthesis A clinical and doppler echocardiographic follow-up. J Thorac Cardiovasc Surg. 1993;106(4):599–608.

Thevenet A, Albat B. Long term follow-up of 292 patients after valve replacement with Omnicarbon prosthetic valve. J Heart Valve Dis. 1995;4(6):634–9.

Abe T, Kamata K, Komatsu K, et al. Ten years' experience of aortic valve replacement with the Omnicarbon valve prosthesis. Ann Thorac Surg. 1996;61(4):1182–7.

Iguro Y, Moriyama Y, Yamashita M, et al. Clinical experience of 473 patients with the Omnicarbon prosthetic heart valve. J Heart Valve Dis. 1999;8(6):674–9.

Torregrosa S, Gomez-Plana J, Valera FJ, et al. Long-term clinical experience with the Omnicarbon prosthetic valve. Ann Thorac Surg. 1999;68:881–6.

Misawa Y, Fuse K, Saito T, et al. Fourteen year experience with the Omnicarbon prosthetic heart valve. ASAIO J. 2001;47(6):677–82.

Kurumisawa S, Kaminiski Y, Muraoka A, Misawa Y. J Cardiothorac Surg. 2016;11:40–4.

Blot WJ, Ibrahim M, Ivey TD, Acheson DE, Brookmeyer R, Weyman A, Defauw J, Smith JK, Harrison D. Twenty-five-year experience with the Björk-Shiley convexo-concave heart valve: a continuing clinical concern. Circulation. 2005;111:2850–7.

Blackstone EH. Could it happen again? The Björk-Shiley convexo-concave heart valve story. Circulation. 2005;111:2717–9.

Björk VO, Lindblom D. The Monostrut Björk-Shiley heart valve. J Am Coll Cardiol. 1985;6:11142–8.

Meir B. The Designer of Faulty Heart Valve Seeks Redemption in New Device. The New York Times, April 17; 1990.

Bodnar E, Frater R, editors. Replacement cardiac valves. Pergamon Press; 1991.

Ibid, Bokros JC et al. The durability of mechanical heart valve replacements: past experience and current trends, Chapter 2, p. 21–48.

Ibid, Schoen FJ. Materials considerations for improved cardiac valve prostheses, Chapter 15.

Bokros JC. Carbon in medical devices (The Pettinos lecture). Carbon. 1977;15:355–71.

Possis Z US Patent No 4,078,268, Heart Valve Prosthesis, issued March 14, 1978.

Ma Ling: Studies on pyrolytic carbon for biomedical applications. A dissertation submitted in partial satisfaction of the requirements for the degree Doctor of Philosophy in Materials Science and Engineering. Approved Aly H. Shabaik, Jenn Ming Yang, John M. Christie and George Sines, Committee Chair. University of California Los Angeles. 1997.

Cook SD, Thomas KA, Kester MA. Wear characteristics of the canine acetabulum against different femoral prostheses. J Bone Joint Surg. 1989;71(B):189–97.

Stanley J, Klawitter JJ, More R. Replacing joints with pyrolytic carbon. In: Revell P, editor. Joint replacement technology. Cambridge/Boca Raton/Boston/New York/Washington, DC: Woodhead Publishing/CRC Press; 2008.

Klawitter JJ, Patton J, More R, Peter N, Podnos E, Ross M. In vitro comparison of wear characteristics of pyrocarbon and metal on bone: shoulder hemiarthroplasty. Shoulder Elbow. 2018;12.

Bokros JC, Emken MR, Akins RJ, et al. European Patent Specification 0055406, filed 071281 and issued 270385.

Podesser BK, Khuenl-Brady G, Eigenbauer E, et al. Long-term results of heart valve replacement with the Edwards Duromedics bileaflet prosthesis: a prospective ten-year clinical follow-up. J Thorac Cardiovasc Surg. 1998;115(5):1121–9.

Johansen P. Mechanical heart valve cavitation. Expert Rev Med Dev. 2004;1(1):95–104.

Andersen TS, Johnasen P, Christensen DO, et al. Interoperative and postoperative evaluation of cavitation in mechanical heart valve patient. Ann Thorac Surg. 2006;81:34–41.

Kent JN, Bokros JC. Pyrolytic carbon and carbon coated dental implants. Dental Clin North Am. 1980;24(3):465.

Bokros JC, Akins RJ, Shim HS, et al. Prostheses made of carbon. Chem Technol. 1977;7:40–9.

Bokros JC. US Patent No 4,692,165, Heart Valve, issued September 8, 1987.

Bokros JC. US Patent No 4,689,046, Heart Valve Prosthesis, issued August 25, 1987.

Stupka JC, Bokros JC, Emken MR, Haubold AD, Peters TS. US Patent No 5,192,309, Prosthetic Heart Valve, issued March 9, 1993.

Accuntius J, Wilde D. US Patent No 5,284,676, Deposition of pyrolytic carbon in fluid bed, issued February 8, 1994.

Ely JL, Haubold AD, Bokros JC, et al. US Patent No 5,514,410, Pyrocarbon and process for depositing pyrocarbon coatings, issued May 7, 1996.

Ely J, Emken M, Accuntius J, et al. Pure pyrolytic carbon; preparation and properties of a new material, On-X® carbon, for mechanical heart valve prostheses. J Heart Valve Dis. 1998;7:626–32.

Bokros JC, Ely JL, Emken MR, et al. US Patent No 5,545,216, Prosthetic heart valve with improved blood flow , issued August 18, 1996.

Bokros JC, Ely JL, Emken MR, et al. US Patent No 5,641,324, Prosthetic heart valve with improved blood flow, issued June 24, 1997.

Bokros JC, Ely JL, Emken MR, et al. US Patent No 5,772,694, Prosthetic heart valve with improved blood flow, issued June 30, 1998.

Bokros JC, Ely JL, Emken MR, et al. US Patent No 5,908,452, Prosthetic heart valve with improved blood flow, issued June 1, 1999.

Swanson WM, Clark RD. Dimensions and geometric relationships of the human aortic valve as a function of pressure. Circ Res. 1974;35:871–82.

Dembitski G. Personal communication, November 1995.

Scotten LN, Siegel R. Importance of shear in prosthetic valve closure dynamics. J Heart Valve Dis. 2011;20:664–72.

Emken MR, Seeley M, Wilde DS. US Patent No 5,305,554, Moisture control in vibratory mass finishing systems, issued April 26, 1994.

Wilde DS, Hightower BF, Accuntius JA. US Patent No 5,332,337, Particle feeding device and method for pyrolytic carbon coaters, issued July 26, 1994.

Waits CT, Stupka JC, Peters TS, Schwartz AS. US Patent No 5,336,259, Assembly of heart valve by installing occluder in annular valve body, issued August 9, 1994.

Laczkovics A, Heidt M, Oelert H, et al. Early experience with the On-X prosthetic heart valve. J Heart Valve Dis. 2001;10:94–9.

Carpentier A, Lemaigre G, Robert L, et al. Biological factors affecting long-term results of valvular heterografts. J Thorac Cardiovasc Surg. 1969;58:457–83.

Summary of safety and effectiveness for On-X 2001 aortic, FDA PMA P000037, May 30, 2001 and March 6, 2002 European Primary Trial Updated May 31, 2003.

Summary of safety and effectiveness for On-X 2002 mitral, FDA PMA P000037/S1, May 30, 2001 and March 6, 2002 European Primary Trial Updated May 31, 2003.

Zilla P, Brink J, Human P, Bezuidenhout D. Prosthetic heart valves: Catering for the few. Biomaterials. 2008;29(4):385–406.

Leonardo da Vinci, An Artabras Book. Reynal and Company in association with William Morrow and Company, New York. All rights reserved under International and Pan-American Copyright Conventions and copyright by Instituto Geografico De Agostini, Novara, Italy.

Leonardo da Vinci on the human body by Charles D O'Malley and JB de CM Saunders. Wings Books, New York. Avenel Random House ISBN 0-517-38105-2

Kilner PJ, Yang GZ, Mohiaddin RH, et al. Helical and retrograde secondary flow patterns in the aortic arch studied by three-directional magnetic resonance velocity mapping. Circulation. 1993;88:2235–47.

Marinelli R, Fuerst B, Vander H, et al. The Heart is not a pump: a refutation of the pressure propulsion premise of heart function. "Frontier Perspectives" the journal of the Center for Frontier Sciences at Temple University, fall-winter 1995 (Volume 5, #1).

Stonebridge PA, Buckley C, Thompson D, et al. Non-spiral and spiral flow patterns; in-vitro observations using magnetic resonance imaging and computational fluid dynamic modeling. Int Angiol. 2004;23(3):276–83.

Kilner PJ, Yang GZ, Wilkes AJ, Mohladdin RH, Firmin DN, Yacoub MH. Nature. 2000;404(6779):759–761.

Körtke H, Körfer R. International normalized ratio self-management after mechanical heart valve replacement: is an early start advantageous? Ann Thorac Surg. 2001;72:44–8.

Kinsley RH, Colsen PR, Antunes MJ. Medtronic-Hall replacement in a third-world population group. Thorac Cardiovasc Surg. 1983;31(11):69.

Meuris B, Flameng W. Performance of bileaflet heart valve prostheses in a new animal model presented at the 6th Annual Hilton Head Workshop of prosthetic heart valve. March 6–10, 2002.

Cannegieter SC, Rosendall FR, Briet B. Thomboembolic and bleeding complications in patients with mechanical heart valve prostheses. Circulation. 1994;89:635–64.

Butchart EG. Prosthetic heart valves, chapter 22. In: Verstracete M, Fuster V, Topol EJ, editors. Cardiovascular thrombosis: thrombocardiology and thromboneurology. 2nd ed. Philadelphia: Lippincott-Raven Publishers; 1998. p. 395–414.

Bockeria LA, Gorodkov AJ, Dorofeev AV, et al. Left ventricular geometry reconstruction in ischemic cardiomyopathy patients with predominantly hypokinetic left ventricle. Eur J Cardiothorac Surg Suppl. 2006:S251–8.

Pasipoularides A. Heart's vortex. Shelton: Peoples Medical Publishing House; 2010.

Coghlan C, Hoffman J. Eur J Cardiothoracic Surg Suppl. 2006;295:S4–S17.

Markl M, Draney MT, Miller DC, et al. Time-resolved three-dimensional magnetic resonance velocity mapping of aortic flow in healthy volunteers and patients after valve-sparing aortic root replacement. J Thorac Cardiovasc Surg. 2005;130:456–63.

Bakhtiary F, Schiemann M, Dzemali O, et al. Impact of patient-prosthesis mismatch and aortic valve design on coronary flow reserve after aortic valve replacement J Am Coll Cardiol 2007;49(7):790-796

Burnett C. Comparisons of valve performance, FDA submission data Medtronic Freestyle®Aortic Root Prostheses Summary of safety and effectiveness data submitted to the United States Food and Drug Administration. PMA P97 0031. Approval date November 26, 1997. On-X Life Technologies, Inc. (technical report)

Badhwar V, Wei LM, Cook CC, et al. Robotic aortic valve replacement. J Thorac Cardiovasc Surg. 2021;161(5):1753–9.

Bokros JC, Stupka JC, Waits CT. U.S. Patent 8,303,652 B2, Heart valve inserter, issued November 6, 2012.

Ruyra-Baliarda FJ, Gatell J, Ferre J, Southard FD, Bokros JC, Correa E, Poehlmann J. US Patent No 9,314,333 B2, Heart Valve Sewing Cuff, issued April. 19, 2016.

Bellhouse BJ, Bellhouse FH. Mechanism of closure of the aortic valve. Nature. 1968;217:86–7.

Bellhouse BJ, Bellhouse FH. Fluid mechanics of the mitral valve. Nature. 1969;224:615–6.

Puskas J, Gerdisch M, Nichols D. Reduced anticoagulation after mechanical aortic valve replacement: interim results from the prospective randomized On-X valve anticoagulation clinical trial randomized Food and Drug Administration investigational device exemption trial. J Thorac Cardiovasc Surg. 2014;147:1202–10.

Malaisrie SC, Barker, et al. Restoration of physiologic flow in the ascending aorta after mechanical aortic valve replacement. Poster from Northwestern University Feinberg School of Medicine presented at Proceedings of the Heart Valve Society, 2015, Monaco.

Von Knobelsdorff-Brenkenhoff F, et al. Blood flow characteristics in the ascending aorta after aortic valve replacement—a pilot study using 4D-flow MRI. Int J Cardiol. 2014;170(3):426–33.

Keller EJ, Malaisrie SC, Krose J, McCarthy PM, Carr JC, Markl M, Barker AJ, Collins JD. Reduction of aberrant aortic haemodynamics following aortic root replacement with a mechanical valve conduit. Interact Cardio Vas Thorac Surg. 2016; https://doi.org/10.1093/icvts/ivw173.

Williams MA, Van Riet S, Jason M. The On-X heart valve at 15 years in a poorly anticoagulated population. Presented at the annual meeting of the Asian Society for Cardiovascular and Thoracic Surgery in Seoul, South Korea. March 2017.

Index